The African Marine Litter Outlook

Thomas Maes · Fiona Preston-Whyte
Editors

The African Marine Litter Outlook

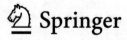

Editors
Thomas Maes
GRID-Arendal
Arendal, Norway

Fiona Preston-Whyte
GRID-Arendal
Arendal, Norway

ISBN 978-3-031-08628-1   ISBN 978-3-031-08626-7 (eBook)
https://doi.org/10.1007/978-3-031-08626-7

This Springer imprint is published by the registered company Springer Nature Switzerland AG
The registered company address is: Gewerbestrasse 11, 6330 Cham, Switzerland

"I dedicate this book to Arthur and Robyn Stella."
- Thomas

Foreword

After years of fairly slow development, Africa is now undergoing rapid economic growth. This rapid growth does have some risks; in the absence of adequate waste management, Africa is becoming a hotspot for plastic pollution.

Globally, plastic production and consumption increased exponentially over the last decades. Although it is a versatile, useful and convenient product, it has become an urgent environmental issue. The rapidly increasing production of plastic and lack of capacity to deal with it, especially single use, disposable items, is leading to a disaster that will engulf the entire world. Plastic pollution is, however, most visible in developing nations, where garbage collection systems are often ineffective or non-existent. Despite worldwide initiatives and efforts, the amount of plastic in the ocean has been estimated to be several hundred million tonnes and the amount of plastic waste entering aquatic ecosystems is increasing at an alarming rate. As a result, plastic pollution is a global concern as well as a significant problem in Africa. Plastic litter taints African capital cities, fresh water, terrestrial environments and Africa's oceans in ever-increasing quantities.

With the growing catastrophe of plastic pollution threatening wildlife, livelihoods, economies and potentially human health, actions at different levels must be accelerated. There is currently, however, no comprehensive assessment of the quantities of plastic present in the environment. In Africa, efforts have been made at the regional, sub-regional and national levels to promote science in order to better understand plastics volumes, types, hotspots and fluxes, among other things. A methodology for monitoring plastic pollution has been developed in some African countries as part of the response. Resolutions and protocols as well as national and regional action plans have also been adopted.

The African Marine Litter Outlook provides a detailed overview of marine litter from the African perspective. Written by experts based in Africa and from around the world, it is an authoritative work founded upon the most up-to-date science. While containing detailed scientific information, this book provides a sound knowledge base for policy-makers, NGOs and the broader public.

Nevertheless, we still need to better understand the cycling of plastic pollution in the ocean, with many knowledge gaps to be filled on sources, degradation and impacts, in all parts of the world and particular in Africa. This will support the United Nations Environment Assembly (UNEA) resolution to end plastic pollution. From the African perspective, the new internationally legally binding instrument is welcomed as it will strengthen a comprehensive approach to address the full life cycle of plastics, from production and design to waste prevention and management. It addresses the need for an international framework to share costs and benefits across the global value chain and stimulate holistic action, that attracts the interest of a number of organisations, in order to combat the transboundary nature of plastic pollution. Each African country is best positioned to understand its own national circumstances, including its stakeholder activities, related to addressing plastic pollution. The new global agreement to stop plastic pollution will thus have to be integrated into existing ocean and coastal management, policy and governance structures.

Most likely, Africa will be able to leapfrog to advanced solutions without having to follow the path traditionally taken by industrialised countries. Although the Global Plastic Treaty is only in its infancy, and much more work needs to be done to implement it successfully, it holds great promise, especially when combined with the strengths and versatility of the African continent.

Dr. Peter T. Harris
Managing Director, GRID-Arendal
Arendal, Norway

Paul A. Lamin
Deputy Director, Natural Resources Management Department
at Environment Protection Agency
Freetown, Sierra Leone

Dr. Francois Galgani
Head of the Laboratory Environment and Resources
IFREMER
Corsica, France

Preface

The African Marine Litter Outlook provides an overview of marine litter from the African perspective. This regional perspective includes a comprehensive collection of the knowledge of marine litter in Africa, including current knowledge, future projections, and potential policy solutions.

The outlook gives an overview of Africa's marine environment, and its importance and the threats it faces (Chap. 1), covers what is known and unknown about Marine Litter Sources and Distribution Pathways (Chap. 2), Impacts and Threats of Marine Litter to the African Continent (Chap. 3), the Legal and Policy Frameworks Relevant to Marine Litter (Chap. 4) and The Way Forward for Africa (Chap. 5).

Chapter 1 takes the reader on a journey on the importance of the ocean, including the value of the Blue Economy. The pressures facing the ocean are introduced with a focus on marine litter and the concept of threat multipliers. These themes are covered from a global perspective before focusing on Africa. The reasons why marine litter, specifically plastic pollution, is a growing threat in Africa are covered—including the regional and global aspects driving this increasing issue. The uncertainties created by COVID-19 and its effects on future projections on marine litter are also discussed in this chapter.

It is important to identify marine litter sources and distribution pathways for the development of targeted and effective interventions and strategies. These have been relatively less researched on the African continent. Chapter 2 reviews quantification studies that have been conducted across the African continent, highlighting knowledge and data gaps that need to be addressed. Current published studies are isolated to a few selected countries with the majority conducted in South Africa. Multiple land and sea-based sources are identified including direct littering and dumping, domestic and industrial wastewater, shipping as well as fishing and aquaculture industries. Beach litter surveys are the most common quantification and monitoring technique employed which may be attributed to the ease of access to beaches and unsophisticated requirements in comparison to other methods. Relatively few marine litter studies have been conducted in freshwater environments and urban infrastructures such as wastewater treatment plants and drainage systems.

Chapter 3 discusses the impacts of marine litter in Africa. Marine litter has environmental, social, economic and human health impacts. Though these topics are discussed separately, they cannot be considered in isolation but are rather interconnected. The size, distribution and quantities of marine litter determine its impact. The emphasis of this chapter is on the coastal countries of the African continent. Compared to the quantitative studies in Chap. 2, there are even fewer studies focusing on the impacts of marine litter in Africa, and so where information from Africa is missing, relevant knowledge from around the world is utilised to infer possible impacts.

Chapter 4 provides a synopsis of the regional and international legal frameworks in relation to marine litter. These instruments set the obligations, guidance and support for national action. The countries that are participating in the various instruments are listed. The application of the different policy instruments to marine litter varies. Compliance at a national level needs to occur in relation to regional and international instruments; this chapter summarises the duties established in Africa, with a summary of the barriers and drivers of effective implementation of national measures.

Chapter 5 provides a summary of the above chapters, a discussion of the main findings, and concludes with suggested ways forward. It gives an overview of recommendations and potential actions which are vital to tackle the issue of marine litter across the African continent. Given the diversity of the African continent, there is a need to develop a decision framework for local, national and regional actions to feed into global commitments. This will assist African nations in implementing the best measures for their unique social and economic situations. Each country is best placed to appreciate its national solutions and limitations. This includes stakeholder involvement and financial and technical capacity needs for addressing plastic pollution issues.

Arendal, Norway   Thomas Maes
                                                          Fiona Preston-Whyte

Acknowledgements

We are grateful to the Government of Norway for their financial support received through the Norwegian Agency for Development Cooperation, Norad.

We would like to acknowledge the invaluable support of the reviewers whose expert knowledge, time and comments helped improve the quality of this book. The following reviewers' have peer-reviewed the chapters:

Clever Mafuta (GRID-Arendal, Arendal, Norway)

Morten Sorensen (GRID-Arendal, Arendal, Norway)

Peter Ryan (FitzPatrick Institute of African Ornithology, University of Cape Town, Cape Town, South Africa)

Anham Salyani (GEMS/Water at United Nations Environment Programme, Kenya)

Salieu Sankoh (Research Fellow at Fourah Bay College—University of Sierra Leone, West Africa Regional Fisheries Programme (WARFP), Ministry of Fisheries and Marine Resources (MFMR), Freetown, Sierra Leone)

Tony Ribbink (Sustainable Seas Trust, Gqeberha, South Africa)

Elvis Okoffo (The University of Queensland, Queensland Alliance for Environmental Health Science (QAEHS), Queensland, Australia)

Jost Dittkirst and the Plastic Task Force of the Secretariat of the Basel, Rotterdam and Stockholm Conventions (UN Environment Programme)

Abdoulaye Diagana (UN ENVIRONMENT/Abidjan Convention Secretariat, Abidjan, Côte d'Ivoire)

Alison Amoussou (Abidjan Convention Secretariat, Abidjan, Côte d'Ivoire)

We would like to acknowledge the superb work provided by Nieves López and Federico Labanti (Studio Atlantis) in turning our words and research into figures and illustrations.

Contents

Contributors

Arabi Sumaiya Department of Forestry, Fisheries and the Environment, Cape Town, South Africa

Chitaka Takunda Yeukai Department of Social Work, University of the Western Cape, Cape Town, South Africa

Maes Thomas GRID-Arendal, Arendal, Norway

Manyara Peter Marine Plastics and Coastal Communities (MARPLASTICCs), IUCN-International Union for Conservation of Nature, Eastern and Southern Africa Region, Nairobi, Kenya

Neehaul Yashvin Reef Conservation, Pereybere, Mauritius

Nel Holly Astrid School of Geography, Earth and Environmental Sciences, University of Birmingham, Birmingham, UK

Onianwa Percy Chuks Basel Convention Coordinating Centre for the African Region (in Nigeria), University of Ibadan, Ibadan, Nigeria;
Department of Chemistry, University of Ibadan, Ibadan, Nigeria

Preston-Whyte Fiona GRID-Arendal, Arendal, Norway

Raubenheimer Karen Australian National Center for Ocean Resources and Security (ANCORS), University of Wollongong, Wollongong, Australia

Sadan Zaynab WWF South Africa, Circular Plastics Economy Programme, Cape Town, South Africa

Sparks Conrad Department of Conservation and Marine Sciences, Cape Peninsula University of Technology, Cape Town, South Africa

Abbreviations

AIM	Africa's Integrated Maritime
ALDFG	Abandoned, Lost, or otherwise Discarded Fishing Gear
AMCEN	The African Ministerial Conference on the Environment
AMD	Africa's Maritime Domain
AMWN	African Marine Waste Network
ATR-FTIR	Attenuated Total Reflection-Fourier Transform Infrared
BCRCs-SCRCs	The Basel and Stockholm conventions benefit from a network of 23 Regional and Coordinating Centres for Capacity Building and Technology Transfer
CO_2	Carbon Dioxide
COP	Conference of the Parties
DSC	Differential Scanning Calorimetry
EAC	The East African Community
ECOWAS	The Economic Community of West African States
EDs	Endocrine Disruptors
FAO	The Food and Agriculture Organization
G7	Group of Seven
G20	Group of Twenty
GDP	Gross Domestic Product
GPA	Global Programme of Action for the Protection of the Marine Environment from Land-based Activities
GPML	Global Partnership on Marine Litter
IUCN	The International Union for Conservation of Nature
km^{-2}	Kilometre squared
LME	Large Marine Ecosystems
MEA	Millennium Ecosystem Assessment
mm	Millimetre
MT	Metric Tonnes
Norad	Norwegian Agency for Development Cooperation
PE	Polyethylene
PET	Polyethylene Terephthalate

PEVA	Polyethylene Vinyl Acetate
POPs	Persistent Organic Pollutants
PP	Polypropylene
PS	Polystyrene
PU	polyurethane
PVC	Polyvinyl Chloride
SADC	The Southern African Development Community
SCP	Sustainable Consumption and Production
SDGs	Sustainable Development Goals
SEAFO	South East Atlantic Fisheries Organisation
SGP	Small Grants Programme
SIDS	Small Island Developing States
TLC	The Litterboom Project
UNCLOS	The United Nations Convention on the Law of the Sea
UNEP	The United Nations Environment Programme
US$	United States Dollar
WIO	Western Indian Ocean
WIOMSA	Western Indian Ocean Marine Science Association
WIO-RAPMaLi	Western Indian Ocean Regional Action Plan on Marine Litter
WWF	The World Wide Fund for Nature
WWTPs	Wastewater Treatment Plants
μm	Micrometre

Chapter 1
Introduction to Marine Litter in Africa

Fiona Preston-Whyte and Thomas Maes

Summary What to expect from the African Marine Litter Outlook? The African Marine Litter Outlook provides an overview of marine litter from the African perspective. The Outlook covers: Marine Litter Sources and Distribution Pathways (Chap. 2), Impacts and Threats of Marine Litter in African Seas (Chap. 3), Legal and Policy Frameworks to address Marine Litter through Improved Livelihoods (Chap. 4), and The Way Forward, Building up from on-the-Ground Innovation (Chap. 5). This chapter provides the context for marine litter in Africa from a global and regional standpoint. This chapter introduces the concept of marine litter, the importance of the ocean, and the value of the Blue Economy in Africa. The uncertainties created by COVID-19 and its effects on future projections of marine litter are also summarised in this chapter.

Keywords Africa · Blue Economy · Waste · Marine litter

1.1 Introduction

The ocean is of great importance to earth, not just to coastal nations but also to landlocked communities and countries. The ocean regulates our planet. It produces vast amounts of the oxygen we breathe and acts as a global climate control system by absorbing, storing, and releasing heat and gasses. It is a source of food and essential nutrients such as iodine. It provides the backbone of global transport and trade. Our ocean provides critical economic opportunities and sustainable industries, and contributes to recreation and mental well-being. Covering >70% of the earth's surface (Kaiser et al., 2005), it is not surprising that the ocean is integral to supporting life on earth, providing us with basic necessities whilst regulating our blue planet.

The coastal and marine environment is crucial for the livelihoods of many inland and coastal communities. In 2005, approximately 2.2 billion people lived within 100 km of coastline. The rapid urbanisation of these areas is causing this figure to

F. Preston-Whyte · T. Maes (✉)
GRID-Arendal, Teaterplassen 3, 4836 Arendal, Norway
e-mail: thomas.maes@grida.no

© The Author(s) 2023
T. Maes and F. Preston-Whyte (eds.), *The African Marine Litter Outlook*,
https://doi.org/10.1007/978-3-031-08626-7_1

double by 2025 (Kaiser et al., 2005). The average population density along coastlines changes rapidly around the globe. On average, globally, the population density is 80 people per km^2. However, this rises to 1000 people per km^2 in countries such as Egypt and Bangladesh (Kaiser et al., 2005). Many coastal inhabitants depend directly on marine resources for their subsistence or income. More than 3 billion people's livelihoods (>40% of the global population [World Bank, 2019]) depend directly on coastal and marine biodiversity, whilst the maritime fishing sector, directly or indirectly, employs more than 200 million people (United Nations, 2021).

The economic potential of the ocean is of global importance. It goes further than the fishing industry. The ocean supports a whole range of maritime activities such as shipyards, marine terminals, aquaculture, seafood processing, commercial diving, and marine transportation. The ocean's contribution to the global economy was predicted, in a pre-COVID-19 projection, to double from US $1.5 trillion in 2010 to US $3 trillion by 2030 (OECD, 2016). The Gross Domestic Product (GDP) of the ocean is of a similar order, estimated at US $2.5 trillion per year (gross marine product) (Hoegh-Guldberg et al., 2015), making the global ocean the words 8th largest economy (World Economic Forum, 2020).

The ocean's economic importance is recognised through the Blue Economy. The Blue Economy is the "sustainable use of ocean resources for economic growth, improved livelihoods and jobs whilst preserving the health of ocean ecosystem." The Blue Economy recognises the need to enhance economic development by exploiting marine resources in a sustainable and regenerative way that conserves life-sustaining marine environments. When approached as the Blue Economy, the maritime industries and activities promote the preservation or improvement of livelihoods, social inclusion and economic growth, whilst ensuring environmental sustainability (United Nations, 2019). The ocean's importance reaches far beyond economic and subsistence. The ocean allows for life on this earth as we know it.

The ocean is a climate regulator and buffer to the current anthropogenic pressures, placing the ocean under the increasing strain of ocean acidification, warming temperatures, decreasing oxygen, and sea-level rise (Bindoff et al., 2019). The diversity of species found in our ocean offer great potential for treatments to combat illness and improve our quality of life. The ocean provides physiological comfort to humans, enhances health and wellbeing, and supports the development of self-efficacy and resilience (Costello et al., 2019). Important in times of increased distress, as seen during the COVID-19 pandemic. A recent study suggests a potential buffer effect of residential proximity to the coast against negative psychological consequences of the COVID-19 pandemic, further supporting the notion that the coast has a positive impact on wellbeing (Severin et al., 2021).

Our ocean is in peril. Overfishing has had a severe impact on fish stocks and ecosystems. Habitats have been degraded, and increasing pollution and climate change further deteriorates entire marine environments. Plastic pollution has entered the ocean by millions of tonnes in the last decades. These threats act as accumulative threats, all of which contribute to rapid biodiversity loss.

Although international conventions provide protection through agreed global cooperation, only 2% of the ocean is protected by marine reserves (Sala et al., 2018). These international conventions include but are not limited to: the international Convention on the Prevention of Marine Pollution by Dumping of Wastes and Other Matter, 1972 (London Convention) and The International Convention for the Prevention of Pollution from Ships, 1973 (MARPOL) and their associated annexes and protocols. Additional regional governance occurs. The seas surrounding Africa are governed through the Barcelona Convention, the Nairobi Convention, the Jeddah Convention and the Abidjan Convention, including its additional protocols on pollution from land-based sources, integrated coastal zone management, sustainable mangrove management, environmental standards and guidelines for offshore oil and gas activities and policy on integrated ocean management. See Chap. 4 and associated figures for details on relevant ocean governance frameworks and their geographical coverage.

The importance of the ocean is emphasised by its role as a climate regulator, a source of food and Blue Economies, all of which have global and regional scales and importance. Due to our physiological, ecological, economic, and psychological dependency on a healthy ocean, ocean protection forms a fundamental part of the United Nations' Sustainable Development Goals (SDGs). In October 2021, the UN Human Rights Council in Geneva recognised access to a sustainable and healthy environment as a universal right. Inclusion of the ocean in the SDGs (SDG14–"Life below water") provides global recognition of its importance, and consensus and incentives for protection.

1.2   The Threats the Ocean Faces

The ocean faces cumulative threats from climate change, biodiversity loss, and pollution. With pollution covering chemical pollution, such as nutrient enrichment leading to eutrophication, marine litter, and physical pollution such as the input of light, noise, and heat. The focus of this African Marine Litter Outlook is on marine litter; however, the pollution threats to the ocean are interlinked. They should be considered in relation to each other and in relation to other cumulative pressures on the ocean such as climate change and biodiversity loss.

1.2.1   Marine Litter

Marine litter is "*any persistent, manufactured or processed solid material discarded, disposed of or abandoned in the marine and coastal environment*" (UNEP, 2006). Though often stated that 80% of marine litter is from land-based sources (GESAMP, 1991; Sheavly, 2005), this number is variable and depends on the location and its pressures. Coastal areas close to urban centres are dominated by litter from land-based litter sources (Ryan et al., 2018). However, the open ocean, islands, and more remote locations show higher proportions of litter from sea-based sources (Ryan, 2020; Ryan et al., 2019). Sea-based sources originate from a wide range of maritime activities, including fishing and shipping related amongst other types of litter. Jambeck et al. (2015) noted that this 80% figure *"is not well substantiated and does not inform the total mass of debris entering the marine environment from land-based sources."* The term marine litter encompasses a wide range of materials and sizes (from mega to nano) spread across a variety of compartments (beaches, seafloor, water surface, and water column, etc.), the largest proportion (61–87%) of marine litter consists of plastic (Barboza et al., 2019; Tekman et al., 2019). This percentage indicates weight, irrespective of units, whilst excluding microplastics.

Box 1.1: Size Fractionate and Definitions of Litter

Plastic litter is defined through size classes as macro (>25 mm), meso (5–25 mm), micro (1 μm–5 mm) and nanoplastics (<1 μm) (GESAMP, 2019). Microplastic enters and forms in the environment through primary and secondary sources, respectively. Primary sources are direct inputs such as plastic nurdles, microbeads from personal care products (Carr et al., 2016), and industry abrasives (GESAMP, 2015). Secondary sources of micro and nanoplastics include: industry abrasives (Eunomia, 2017), textile fibres (Browne et al., 2011; Napper & Thompson, 2016), tyre dust as well as fragmentation and degradation of macro and mesoplastics (5–25 mm) (GESAMP, 2019).

A Global Perspective of Marine Litter

Following their commercial development in the 1930s (Jambeck et al., 2015), plastic production only grew substantially from the 1950s, with an increase in single-use plastic consumption items and resulting "throw-away" culture in the last two decades—seeing half of the plastic produced since the 1950s was made between 2002 and 2015 (Geyer et al., 2017). Geyer et al. (2017) estimated that, as of 2015, 8.3 billion metric tonnes (MT) of virgin plastic had been produced, with approximately 6.3 billion MT of that becoming waste, of which 79% ended up in landfills or the natural environment. Eriksen et al. (2014) estimate that 250,000 tonnes of plastic are afloat in the ocean. This number excludes items made from polymers which are denser than seawater (which account for roughly 40% of global plastic production by mass) as well as less dense polymers that sink due to biofouling (Fazey & Ryan, 2016; Lobelle & Cunliffe, 2011).

Plastic production is increasing. Lebreton and Andrady (2019) estimated that production will double within the next 20 years, and plastic waste production to more than double in the same period (Geyer et al., 2017). In addition to increasing plastic production, further economic investment in the industry in America is projected to accelerate virgin plastic production (Borrelle et al., 2020). Globally, waste management systems are not sufficient to safely dispose of or recycle waste plastic (Velis et al., 2017; Wilson & Velis, 2015), resulting in an inevitable increase in plastic pollution entering the environment (Lau et al., 2020).

Increasing awareness of marine litter has led to numerous and varied attempts to quantify the amount of plastic entering our ocean (UNEP, 2020). Lebreton et al.'s (2017) model estimated the amount of plastic waste currently entering the ocean every year from rivers to be between 1.15 and 2.41 million tonnes. Jambeck et al. (2015) proposed that in 2010, 275 million MT of plastic waste was generated in 192 coastal countries, of which 4.8–12.7 million MT entered the ocean. Jambeck et al. (2015) highlighted that population size and quality of waste management systems largely determine which countries contribute to plastic marine litter. Although these numbers are contested, the scale of the issue is not (Ryan, 2020; Vester & Bouwman, 2020).

Lau et al. (2020) modelled different scenarios with all currently feasible interventions, finding that even with immediate and concerted action, 710 million MT of plastic waste will still cumulatively enter aquatic and terrestrial ecosystems between 2016 and 2040. Borrelle et al. (2020) supported their finding that current efforts are insufficient to tackle the plastic waste problem.

Marine litter is symptomatic of more significant issues. First, resource abuse through the unsustainable design, production, and consumption of single-use items. And second, a lack of service delivery and safe management of waste.

1.2.2 Climate Change

Carbon emissions from anthropogenic activities are causing ocean warming, acidification, and oxygen loss within the marine environment. These changes affect marine organisms, ecosystems (Bindoff et al., 2019), and their services. Marine litter is a threat multiplier to the ocean ecosystems already affected by climate change (UNEP, 2021). Marine litter also contributes to climate change through greenhouse gas emissions (Ford et al., 2022; UNEP, 2021) and by reducing the efficiency of the ocean and their ecosystems in storing CO_2. Similarly, climate change, through sea level rise, storm surges, and flooding are liking to increase the transport of litter into the marine environment.

The Ocean: Carbon Uptake and Storage—Marine Litter's Role as a Threat Multiplier

The ocean contains the most extensive stock of mobile carbon on earth, containing 50 times more than the atmosphere and 10 times more than the stores in plants and soils combined (Sabine et al., 2004). Through both the physical solubility and biological carbon pump, the ocean helps to buffer anthropogenic climate change through CO_2 uptake. The physical solubility pump is the uptake of atmospheric CO_2 in surface waters and its transport into the ocean depths through the sinking of denser water. The biological carbon pump is the uptake of CO_2 by phytoplankton, and its descend to deeper waters through the sinking of organic debris. Furthermore, the Carbon Dioxide and Carbonate System allows living organisms such as shellfish and corals (as well as phytoplankton and zooplankton) to build their shells or carbonated components from calcium carbonate. Thus, removing carbon from the ocean, further leading to a drawdown of CO_2 into the ocean to maintain the carbonate-carbon dioxide balance. Since the industrial revolution, the ocean has acted as a primary net sink (Sabine et al., 2004), taking up 20–30% of anthropomorphic CO_2 in the last two decades (Bindoff et al., 2019). Comparing the ocean carbon uptake to the rainforests in the Congo Basin: every year, the global ocean absorbs almost 8 times more carbon than the Congo rainforest (Gruber et al., 2019; Lewis et al., 2009). However, the ocean is nearing the limit of its ability to provide this buffer as increasing temperatures and ocean acidification threaten the efficiency of these systems (Field et al., 2002; Friedlingstein et al., 2006; Fung et al., 2005).

Though an emerging field of research, marine plastics have been shown to reduce the efficiency of the biological carbon pump by either affecting the survival of zooplankton or by increasing buoyancy of sinking particles (such as faecal pellets or excrement) (Cole et al., 2016; Wieczorek et al., 2019). The biological pump is driven by sinking organic matter, which contains carbon, such as faecal pellets and dead plankton, to the deeper ocean and seafloor, where the carbon becomes trapped. Microplastics reduce feeding, reproductive success, and survival rates in zooplankton (Cole et al., 2015; Lee et al., 2013); this species-level effect may reduce the efficiency of the biological carbon pump. Additionally, when plastic occurs in zooplankton faecal pellets, the pellets become more buoyant, reducing their sinking rates (Cole et al., 2016; Wieczorek et al., 2019). The slower they sink, the more time carbon has to escape back into the upper ocean and atmosphere, thus reducing the efficiency of this sink. Current levels of microplastic ingestion probably have minimal impact on the biological pump. However, under future microplastic concentrations (or in areas with elevated plastic concentrations such as convergent zones), microplastics may have the potential to reduce the efficiency of the biological carbon pump (Wieczorek et al., 2019).

Further effects of plastic on the ocean pumps are theoretical but unknown. These might include interference in light penetration for photosynthesis impacting the biological carbon pump, as well as interfering in the Carbon Dioxide and Carbonate System through smothering of relevant species (e.g., bivalves). Plastic pollution affects microbial biodiversity by altering community composition (Harvey et al., 2020), however, the effects of this are unknown.

Marine Ecosystems and Their Services—Marine Litter's Role as a Threat Multiplier

In addition to the uptake and storage of CO_2 by phytoplankton and ocean waters, coastal marine ecosystems play a crucial role in carbon uptake and storage. Mangrove forests, seagrass meadows, and salt marshes all capture and store substantial amounts of organic carbon through net primary production, burial (in detritus and sediment), and export (Duarte, 2017; Duarte et al., 2013; McLeod et al., 2011). In addition to acting as carbon sinks, these ecosystems, and coral reefs form natural coastal defenses (Temmerman et al., 2013), protecting coastal areas from climate-driven rising sea levels and storm surges (Oppenheimer et al., 2019).

In addition, biomass production and water purification are amongst the most essential ecosystem services delivered by the ocean (Karani & Failler, 2020). Efforts to conserve and restore natural carbon sinks will help reduce the impacts of increases in anthropogenic CO_2 emissions (Mcleod et al., 2011). Blue Carbon and Ecosystem Services offer an opportunity to develop coastal projects to protect these ecosystems and mitigate climate change (Karani & Failler, 2020).

Key coastal ecosystems are exposed to pollution, reducing their ecosystem services. Harris et al. (2021) found that 54% of mangrove forests are within 20 km of a river that discharges substantial plastic pollution (>1 t year^{-1}), compared to 24% of seagrass meadows, 23% of salt marshes, and 17% of coral reefs. Ecosystem services provided by mangrove forests, seagrass meadows, salt marshes, and coral reefs are outlined in Table 1.1.

Coral reefs, and the ecosystem services they provide, are estimated to be worth US \$130,000 ha^{-1} year^{-1} (Diversitas, 2009). Green and Short (2003) estimated the global value of seagrass meadows services to be US \$3.8 trillion, and Costanza et al. (1997) estimated the seagrass meadows to be worth US \$19,000 ha^{-1} year^{-1}. From a biodiversity consideration, mangrove forests and seagrass meadows are two of the most valuable marine habitats in the world, rivalled only by coral reefs in the biodiversity they support (Kaiser et al., 2005).

Mangrove forests and salt marshes trap plastics (Martin et al., 2020; Yao et al., 2019), creating a harsh environment where fragmentation occurs, leading to increased microplastic quantities in their respective biota (Deng et al., 2021; Yao et al., 2019). Seagrass meadows are within close proximity to rivers discharging substantial amounts of plastic pollution (Harris et al., 2021). They grow in naturally sheltered areas, and plastics have been found to settle on their above-ground structure (de Smit et al., 2021; Seng et al., 2020). Seagrass meadows and coral reefs both act as significant microplastic sinks, facilitating the accumulation and burial of microplastics along with sediment, thus removing them from the pelagic food chain (de Smit et al., 2021). Research on the ingestion and impact of these plastics on these ecosystems is in its infancy (Seng et al., 2020). Plastic pollution has been linked with a decline in coral health and increased coral disease such as white syndromes (Lamb et al., 2018; Reichert et al., 2018). Mangrove forest health (tree density, survival, and tree size) is significantly affected by plastic pollution. Plastic

Table 1.1 Ecosystem services provided by mangrove forests, seagrass meadows, salt marshes, and coral reefs (Kaiser et al., 2005; Karani & Failler, 2020; Temmerman et al., 2013)

Ecosystem service	Ecosystem			
	Mangrove forests	Seagrass meadows	Salt marshes	Coral reefs
Carbon sink	x	x	x	
Natural sea defence	x	x	x	x
Biomass production	x	x	x	x
Nursery grounds for fisheries species	x	x		
Reduce land sediment runoff	x	x	x	x
Nutrient input and energy flux with other marine ecosystems	x	x	x	x
Provide food for large, endangered grazers		x		
Human food and fibre source		x		x
Prevents erosion	x	x	x	x
Water purification	x	x	x	x
Mitigate eutrophication		x		
Bind organic pollutants	x	x		
Foraging ground for marine fish	x	x	x	x

directly and indirectly causes mangrove degradation, thus reducing ecological functioning and ecosystem services (Suyadi & Manullang, 2020). Although macroplastics alter seagrass architecture and may prevent vertical rhizome growth (Menicagli et al., 2021), additional effects of plastic pollution on seagrass meadows are largely unknown (Bonanno & Orlando-Bonaca, 2020). Similarly, the impact of plastic pollution on salt marshes is undetermined.

Marine litter provides additional habitats for a range of species. The colonisation of marine litter affects dispersion rates and life-history traits of rafting species. This can lead to increased introduction and colonisation of invasive species; for example, plastics foster the spread of non-native macroalgae in seagrass meadows, thus increasing their vulnerability to invasion.

Marine litter affects the health and growth of seagrass meadows, mangroves, and corals and, therefore can affect the carbon uptake in these ecosystems. Identifying and addressing the sources of marine litter is paramount in mitigating climate change (McLeod et al., 2011).

Climate Change—Plastic Production's Role

The production and management of plastics also contribute to climate change as every stage of the plastic lifecycle releases greenhouse gas emissions (Ford et al., 2022; Shen et al., 2020). The two main forms of energy used by the plastics processing industry are electricity and natural gas. Plastic production includes both direct and indirect greenhouse gas emissions. Direct emissions include the combustion of fuel, such as at a plastics processing facility. Indirect emissions would include fossil-fuel-powered electricity used in plastics processing. With plastic production increasing globally (Geyer et al., 2017), we utilise more fossil fuels in plastic manufacture, thus contributing to climate change. In addition to overuse, the mismanagement of plastic waste and the loss of this resource from a potential circular economy, driving raw material extraction and production, exacerbating climate change (Masnadi et al., 2018; Shen et al., 2020).

Marine Litter—Climate Changes as a Threat Multiplier

Flooding and sea-level rise associated with climate change and global warming will increase the quantities of plastics washed into the ocean every year. In Africa, where poor waste management systems are common, flood water and runoff water from rain are the main pathways through which plastics are introduced into the ocean. Climate change is likely to compound the marine litter issue through increases in sea level, storm events, flooding, and displacement of human settlements leading to issues around the safeguarding and provision of waste management services.

Coastal areas are most vulnerable to the impact and risks associated with sea-level rise in relation to climate change (Lam et al., 2012; World Risk Report, 2018). Global warming and associated climate change and climate variabilities pose huge potentials for disruption of the marine ecosystem globally, with the African continent as no exception. In fact, climate change is, and will, affect the world disproportionately, with Africa being one of the areas disproportionately affected (IPCC, 2019), with climate change further aggravating the physical, biological, social, and economic stress that currently exists in Africa's coastal areas. Climate change could ultimately lead to loss and fragmentation of marine habitats and biodiversity, and especially negative stresses to fishing, aquaculture, food production, and food security in many African coastal cities and settlements. Many African cities are low-lying and are very susceptible to flooding that could become severe in sea-level rise due to climate change. The UN-HABITAT (2008) report has identified many African coastal cities as being at risk in this regard. The coastal settlements of the several small island countries of the African continent are also at high risk. Landfills, or dumpsites positioned near beaches (e.g., Strandfontein and Witsand, Western Cape, South Africa) lie within the zone of projected sea-level rise and are at risk of being breached. In addition, plastic currently buried in beaches will be released into the sea as storm surges and rising sea levels scour coastal areas (Ryan, 2020). Ongoing efforts at mitigation and adaptation to climate change, promoted and supported by international treaties, have achieved very little to mitigate these risks to Africa's coastal cities.

Climate change and its resulting influences on environmental degradation, water stress, and food security impact urbanisation plans (IPCC, 2019; Niang et al., 2014). Waste management systems that are already under pressure through population growth and rapid urbanisation (UNEP, 2018b), will be strained further due to climate change.

1.2.3   Depletion of Fish Stocks

Overfishing and the resulting depletion of fish stocks is a global issue, with the average state of global fish stocks being poor and declining (Costello et al., 2016). Overfishing has ecological, social, and economic effects.

Contributions of Marine Litter to Overfishing

Abandoned, lost, or otherwise discarded fishing gear (ALDFG) is a recognised issue of marine litter from sea-based sources—fisheries specifically. With much of it being synthetic, ALDFG contributes to marine litter, with an estimated 0.6 MT of ALDFG entering the marine environment as litter in 2015 (UNEP, 2018a). Once lost at sea, ALDFG continues to fish. This is referred to as "ghost fishing" and has detrimental impacts on fish stocks and potential impacts on endangered species and benthic environments. The scale of the issue of ghost fishing and its impacts on fish stocks has not been fully quantified. Please see Chap. 3 for more details on the impact of ALDFG.

1.2.4   Pollution

Marine litter and plastic are not the only form of pollution threatening coastal and marine environments; chemical pollution and nutrient enrichment leading to eutrophication are also pollution threats. Chemical pollution covers toxic metals, persistent organic pollutants (POPs), and crude/refined petroleum oil. Nutrients input into rivers and coastal water could result in eutrophication.

Marine litter can act as a threat multiplier to existing chemical and nutrient pollutants. Firstly, by containing additional chemicals. Secondly, by sorbing and transporting chemicals between compartments, areas, and species.

Other Pollutants—Marine Litter's Role as a Threat Multiplier

Plastic has been identified as a vector for toxic chemicals. Plastics contain and leach additives such as colourants, plasticisers, lubricants, and flame retardants into the environment (Rochman et al., 2019).

Plastic particles may also sorb and accumulate chemicals from their surroundings, including POPs and heavy metals (Näkki et al., 2021). These chemicals are then transported with the plastics across environmental compartments where they may be released, or the toxic plastic may be ingested by marine organisms. The net contribution of plastic ingestion to bioaccumulation of contaminants by marine organisms is likely to be small compared to the uptake of contaminants directly from the water itself (Bakir et al., 2012; Koelmans et al., 2016). However, the multi-stressor effect does still need to be considered (see details in Chap. 3).

Other Pollutants—Nutrient Enrichment

Nutrient enrichment of coastal and marine waters is the primary cause of eutrophication that leads to the formation of algal blooms. Eutrophication leads to hypoxic and anoxic conditions in water, extreme turbidity, and a threat to marine life (Malone & Newton, 2020). Nutrient input to the marine environment is primarily derived from land-based sources, mainly through stormwater runoffs from agricultural land where fertilisers are applied. Nutrients are also derived from the discharge of untreated sewage and industrial/domestic wastewater into river courses. As such, there is a similarity with plastic input sources. In Africa, due to the poor state of water and sanitation facilities (Yasin et al., 2010), a significant proportion of the nutrient input originates from sewage disposal. Several eutrophic coastal areas or death zones now affect countries around the African continent, namely Côte d'Ivoire, Egypt, Ghana, Kenya, Mauritius, Morocco, Nigeria, Tanzania, Tunisia, Senegal, and South Africa (Diaz et al., 2011). These might also be hotspots for plastic pollution. Solving the stormwater and wastewater treatment issue would thus reduce pollution from nutrients, sewage, and plastic.

1.3 Africa's Oceanographic Position

As shown in Fig. 1.1a, the African continent is surrounded by the Atlantic Ocean, Indian Ocean, Mediterranean Sea, and the Red Sea. Off the continent's coast are various islands associated with the continent and included in this Outlook. The upwelling linked to the colder Benguela and Canary Currents drive the productivity seen off West Africa. The warmer currents of the east coast of Africa are significant as they bring oceanic water from countries in south-east Asia, which is important for the long-distance drift of marine litter (Duhec et al., 2015; Ryan, 2020; Ryan et al., 2021)–which is covered in detail in Chap. 2.

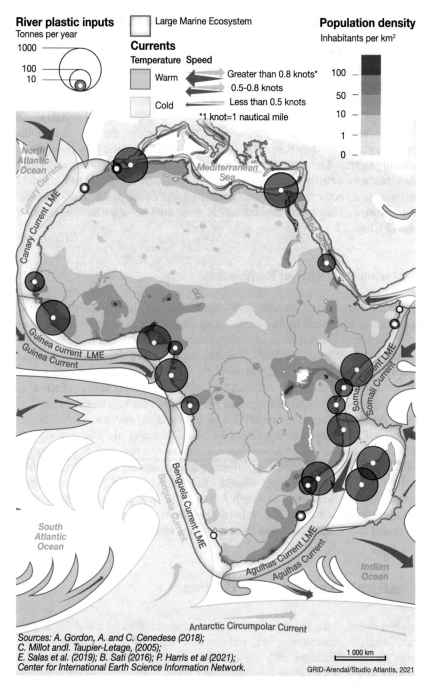

Fig. 1.1 a Africa's oceanographic context showing major currents, LMEs, country population densities, as well as the major river systems and the river's plastic inputs. **b** Africa's key oceanic ecosystems—salt mashes. **c** Africa's key oceanic ecosystems—seagrass meadows. **d** Africa's key oceanic ecosystems—mangrove forests. **e** Africa's key oceanic ecosystems—coral reefs

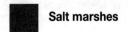 **Salt marshes**

Source: UN Evironment Programme
World Conservation Monitoring Centre.

GRID-Arendal/Studio Atlantis, 2021

Fig. 1.1 (continued)

Seagrass meadows

*Source: UN Evironment Programme
World Conservation Monitoring Centre.*

GRID-Arendal/Studio Atlantis, 2021

Fig. 1.1 (continued)

 **Mangrove forests**

Source: UN Evironment Programme
World Conservation Monitoring Centre.

GRID-Arendal/Studio Atlantis, 2021

Fig. 1.1 (continued)

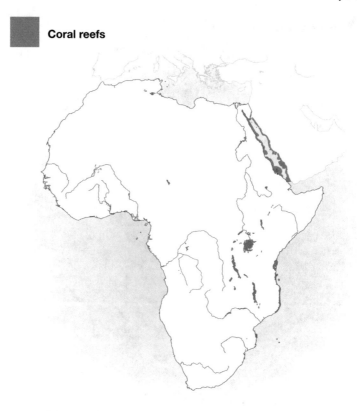

Coral reefs

Source: UN Evironment Programme
World Conservation Monitoring Centre.

GRID-Arendal/Studio Atlantis, 2021

Fig. 1.1 (continued)

Africa has 30,500 km of coastline, with 70% of Africa's 54 countries containing coastlines. The African Large Marine Ecosystems (LMEs) include 33 coastal and 5 Island states (Clarke et al., 2020) and 1.3 billion people as of 2019 (UNDESA, 2019). LMEs are highly biodiverse areas of ocean extending from estuaries to the edge of continental shelves or to the outer boundaries of major currents. Due to the high levels of land-sourced nutrients, these are the world's most productive ocean areas, where most (about 90%) of the world's fish catch is caught. Global LMEs provide essential ecosystem services (US $3 trillion globally) ("LME Hub," 2021). These areas face high levels of degradation due to pollution (including but not limited to marine litter), overfishing, and climate change. See Sects. 1.2.1, 1.2.2, 1.2.3 and 1.2.4 for more detail.

There are 7 LMEs around Africa: Canary Current LME, Mediterranean Sea LME, Red Sea LME, Somali Coastal Current LME, Agulhas Current LME, Benguela Current LME, and Guinea Current LME (Sherman et al., 2011). These LMEs are determined by their currents, most of which are transboundary in nature (Fig. 1.1a). African LMEs are richly endowed with both living and non-living resources including unrivalled natural beauty, and abundant fisheries (Satia, 2016). All of the African LME's have potential for sustainable economic growth (AU-IBAR, 2019a). Governed by strong upwelling systems, the Benguela and Canary currents rank second and third in the world, respectively, in primary productivity (Lutjeharms & Bornman, 2010).

Africa's LMEs contain ecosystems critical to carbon storage and coastal protection, such as, mangroves forests, seagrass meadows, salt marshes, and coral reefs (Fig. 1.1b–e) (see Sect. 1.2.2 for more details). Their current value is estimated to match the average monetary value of carbon uptake and storage of US $130,000 per km^2 of mangrove forests, salt marshes, and seagrass meadows (Karani & Failler, 2020). The Guinea Current LME has some of the world's largest mangrove ecosystems (FAO, 2007; UNEP, 2007), and all the eastern LMEs contain coral reefs (Fig. 1.1a–e).

1.3.1 Africa's Blue Economy

Leaders across Africa recognised the importance of the Blue Economy as an area of future inclusive and sustainable economic growth. The African Union Inter-African Bureau for Animal Resources (AU-IBAR) has developed the Africa Blue Economy Strategy (Karani & Failler, 2020), which outlines the following key sectors (AU-IBAR, 2019a):

i.   Fisheries, aquaculture, and ecosystems conservation,
ii.   Shipping, transportation, and trade,
iii. Sustainable energy, extractive minerals, gas, and innovative industries,
iv.   Environmental sustainability, climate change, and coastal infrastructure,
v.   Governance, Institutions, and social actions.

Inland water masses (e.g., Lake Victoria, Lake Malawi, Lake Tanganyika, etc.) and significant rivers (Fig. 1.1a) are included in the Blue Economy in Africa due to their importance for inland fisheries and transport. The Africa Blue Economy Strategy integrates existing global and African strategies, policies, and initiatives (AU-IBAR, 2019a).

In 2019, Africa had a combined GDP of US $2.6 trillion (International Monetary Fund, 2019). The Blue Economy in Africa in 2018 created 49 million jobs (Fig. 1.2a) and was valued at US $296 billion (11% of the total combined GDP) (Fig. 1.2a, b). Pre-COVID-19 projections estimated these numbers would increase to a value of US $405 billion and 57 million jobs by 2030 (AU-IBAR, 2019a).

African countries have vast LMEs (Satia, 2016), which are attractive to coastal and marine tourism (Karani & Failler, 2020). Thus, within Africa, tourism is the highest value sector of the Blue Economy, both in jobs and monetary terms, now (24 million jobs, US $85 billion) and in the future (28 million jobs, US $135 billion in 2030) (Fig. 1.2a, b).

With large African oil and gas reserves (Satia, 2016), the extractive industry, mining and quarrying, including oil and gas production (Fig. 1.2b) are second after tourism in terms of monetary value (respectively US $56 billion and US $80 billion, in 2018). However, they contribute substantially less towards job creation (Fig. 1.2a) (0.16 million jobs in 2018 and projected to grow to 1.2 million jobs in 2030). Their development can also substantially negatively affect the growth of three of the other sectors of the Blue Economy, that being (i) Fisheries, aquaculture, and ecosystems conservation, (ii) Environmental sustainability, climate change, and coastal infrastructure, and (iii) Governance, Institutions, and social actions, as well as impact climate change.

The African LMEs are some of the most productive globally, with the fisheries sector (Fig. 1.2a, b) employing many people (13 million jobs and 14.7 million jobs in 2018 and 2030, respectively) (AU-IBAR, 2019a). It is noted that (certainly in West Africa) the industrial fisheries are captured mainly by foreign companies, with most of the fish destined for export (Belhabib et al., 2018). African fisheries are overexploited by European and Chinese fleets, with substantial economic losses to Africa. For example, China and Europe pay as little as 4 and 8% of the landed value, respectively, to access West African fishing grounds (Belhabib et al., 2015). Underreporting and illegal practices occur across both fleets. Underreporting impacts local economies and sustainability directly as it hides over-fishing, threatening the long-term sustainability of fishing stocks (Belhabib et al., 2015).

Aquaculture (Fig. 1.2a, b) is currently relatively small, contributing less than 3% globally (Halwart, 2020). The aquaculture industry in Africa is valued (Fig. 1.2b) at US $2.77 billion in 2018 and is projected to expand to US $5.1 billion in 2030 (AU-IBAR, 2019a). Aquaculture employed 1.2 million people in 2018 and is projected to increase to 1.6 million people by 2030 (Fig. 1.2a).

With increasing populations, growing economies (UNEP, 2018b; United Nations, 2020), and trade agreements, port calls, and shipping are expected to grow at a constant rate. Current infrastructure capacity is insufficient to deal with current waste in ports, raising concerns around waste management for future growth

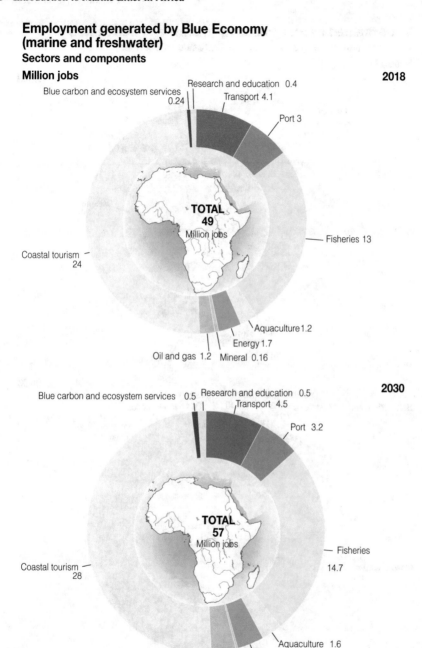

Employment generated by Blue Economy (marine and freshwater)
Sectors and components
Million jobs   **2018**

Blue carbon and ecosystem services 0.24
Research and education 0.4
Transport 4.1
Port 3

TOTAL 49 Million jobs

Fisheries 13

Coastal tourism 24

Aquaculture 1.2
Energy 1.7
Oil and gas 1.2 Mineral 0.16

2030

Blue carbon and ecosystem services 0.5 Research and education 0.5
Transport 4.5
Port 3.2

TOTAL 57 Million jobs

Fisheries 14.7

Coastal tourism 28

Aquaculture 1.6
Energy 2
Oil and gas 1.8
Mineral 0.2

Source: AU-IBAR, 2019a.   GRID-Arendal/Studio Atlantis, 2021

Fig. 1.2 a The value of the Blue Economy (marine and freshwater) in employment terms in 2018 and pre-COVID projection for 2030. **b** The value of the Blue Economy (marine and freshwater) in monitory terms in 2018 and pre-COVID projection for 2030. The size of the graph represents the relative size of Blue Economy

Value created by Blue Economy (marine and freshwater)
Sectors (value added) and Components (value of services)
Billion US$

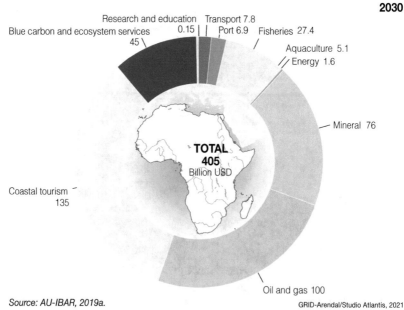

Source: AU-IBAR, 2019a.

GRID-Arendal/Studio Atlantis, 2021

Fig. 1.2 (continued)

(Maes & Preston-Whyte, 2022). Ecosystem services, including blue carbon produced by coastal, marine, and aquatic ecosystems are expected to progressively increase in value as conservation efforts expand. Education and research will follow the same pattern due to a growing demand for knowledge, especially in areas of deep-sea mining, offshore exploration, and climate change mitigation and adaptation (Fig. 1.2a, b) (AU-IBAR, 2019a).

1.4   Marine Litter—A Growing Problem in Africa

Given the present and potential importance of the Blue Economy to Africa, any threat to it, especially to tourism, should be considered a priority. Marine litter is one such threat. The 2019 African Ministerial Conference on the Environment (Durban, South Africa) emphasised the need to address plastic pollution, with all 54 member states supporting a declaration calling for global action on plastic pollution (de Kock et al., 2020). If unchecked, the increase in marine litter will have disastrous consequences on the environment and create socioeconomic development challenges that will impact biodiversity, infrastructure, tourism, and fisheries' livelihoods (Jambeck et al., 2018). Important pressures include population growth, rapid urbanisation, projected economic growth, and increased trade combined with an already constrained public and private sector waste services and infrastructure (Godfrey et al., 2019; UNEP, 2018b) driving increased marine litter inputs.

The current land-based sources of marine litter in Africa are driven by linear product design, population growth, and lack of infrastructure and services. There is increasing dependence on commercially available single-use products which don't consider recyclability or reuse in the design phase. With 3.5% annual growth, Africa's population is growing faster than any other continent (UNEP, 2018b; Wilson et al., 2015). In addition to population growth, Africa has seen rapid urbanisation driven by a changing climate (Henderson et al., 2017), environmental degradation, and socioeconomic factors (Awumbila, 2017). Housing, urban planning, infrastructure for waste (including recycling services) and sanitation services have not managed to keep up with this rapid urbanisation (AfDB et al., 2020; UNEP, 2018b). Thus, this rapid development corresponds with the increase of both macro and microplastics leaking into the environment from the waste stream across Africa (Alimi et al., 2021; Jambeck et al., 2015, 2018; UNEP, 2018a, 2018b). The issues around waste management in Africa are well captured by the African Waste Management Outlook (UNEP, 2018a).

It should be noted that in the context of Africa, the Blue Economy and marine litter, the upstream aquatic environments such as rivers and lakes, and the litter that feeds into them are important. The importance of the upstream aquatic environments is not only limited to the pollution of these water sources but also to the transport of litter through the riverine systems to the sea. Figure 1.1a shows the extensive range of African rivers, many of which are transboundary in nature and pass-through landlocked countries. Inland cities in Africa are often positioned on rivers, and part of their mismanaged waste may be transported to the ocean through rivers (see Chap. 2 for more details). It is noted that this transport may be river-dependent, as comparative research monitoring litter in rivers is in its infancy (UNEP, 2020). Litter also becomes trapped on river banks and in sediments (Weideman et al., 2020). Heavy rains or flooding may see the discharge of this litter into the marine environment (Biermann et al., 2020). The extent of settlement of litter within riverine systems versus transport to the ocean is dependent on the hydrological conditions within the river catchment and is influenced by the climate and weather as well as local sources (Biermann et al., 2020; Moss et al., 2021; Ryan & Perold, 2021; Tramoy et al., 2020; Weideman et al., 2020). The importance of the landlocked African countries in tackling waste management and the resulting marine litter issue highlights the shared responsibility and the need for a regional response.

Africa is not currently managing its waste volumes in environmentally sound nor sustainable ways. The UNEP/IUCN reports highlight this, suggesting that as much as 90% of waste is mismanaged in Kenya, Tanzania, and Mozambique (IUCN et al., 2020a, 2020b, 2021). Future projections raise further concerns regarding Africa's ability to deal with future waste and resulting marine litter, highlighting the need to develop a circular economy within Africa. The future projections which raise these red flags are outlined in the following points:

First, population growth in Africa shows no sign of abating (Wilson et al., 2015). Currently, the African continent has the fastest-growing population with an annual increase of about 2.51%. With a population of 1.34 billion in 2020, pre-COVID-19 predictions by the United Nations estimated that the African population will reach 2.83 billion in 2050, or 40% of the world population, against the current 17% by 2100 (AU-IBAR, 2019a; UNDESA, 2017, 2019). An increase in population is linked to increased waste production. For Sub-Saharan Africa, the annual generation rate of waste was 174 million tonnes in 2016; with it projected to either triple or quadruple by 2050 (Kaza et al., 2018; World Bank, 2018). Whilst, North Africa is expected to double its annual waste generation rates by 2050 (Hoornweg & Bhada-Tata, 2012; Kaza et al., 2018).

Second, economic drivers of urbanisation and migration in Africa are sensitive to climate change impacts (Niang et al., 2014). Rapid urbanisation is projected to continue, with most cities in coastal zones or near river systems (Jambeck et al., 2018). Rapid urbanisation in Africa is such that the number of people living in urban environments are projected to rise from 11.3% in 2010 to 20.2% by 2050 when comparing the percentage of African urban areas to total global urban areas (Awumbila, 2017). Increased urbanisation leads to more waste generation in the urban areas, creating more strain on already underdelivering waste systems. Additionally, rapid urbanisation causes a rise in the cost of land in the cities and housing and an increase in informal settlements. This leads to several "practical solutions," such as in Freetown, Sierra Leone, where solid waste, including plastics, is increasingly being used to reclaim coastal land for the construction of informal settlements and residential houses (Sankoh, 2021, personal communication).

Third, Africa has seen increased economic activity and resulting in increased GDP since the 1980s (continental average), with a pre-COVID-19 growth of 3.7% per year (International Monetary Fund, 2021). As solid waste generation is strongly correlated with gross national income per capita (Hoornweg et al., 2013; Wilson et al., 2015), such projected increases in GDP are likely to be associated with increased waste.

Fourth, Africa's economic development is associated with a growing middle class (Ncube & Lufumpa, 2014; Scharrer et al., 2018), which will likely result in a further increase in spending and associated solid waste generation.

Fifth, increased economic activity and increased trade internationally have seen, and is predicted to see, an increase in shipping and transport. With 3% of the world's volumes, African shipping has a relatively small impact on international trade. However, in the 5 years preceding COVID-19, traffic in African container ports grew 3% faster than global levels (AU-IBAR, 2019b). This means that shipping in African waters increased, and ports were expected to process more waste from ships. Ryan (2020) and Ryan et al. (2021) show, using plastic bottles as an indicator, that despite the existence of MARPOL Annex V, plastic waste is still being dumped from ships. Thus, until adequate enforcement of MARPOL occurs globally, increased shipping is likely to be associated with increased marine litter from ships. However, for MARPOL Annex V to be adequately enforced, there needs to be proper port reception facilities, downstream waste management infrastructure, and port enforcement. Concerningly, a study of the port reception facilities in South Africa shows this is not the case (APWC, 2020). As South Africa has one of the largest economies in Africa and assuming enforcement is linked with economic development, similar lapses in infrastructure can be expected across Africa. Linking in with increased trade, an increase in the extensive packaging used in the shipping of shipped goods and the shipped goods themselves can be observed. Used goods (everything from used electronics, cars, clothes etc.) are imported into Africa as second-hand goods and/or charitable donations (Maes & Preston-Whyte, 2022; UNDESA, 2020). The working order, usability, life span, and quality of these goods is raising increasing concern as an additional waste stream for Africa to deal with (Maes & Preston-Whyte, 2022).

Sixth, in January 2018, China closed its borders to certain imported recyclable materials, including most plastic. Until then, China had been the world's leading importer of plastic waste. This resulted in global turmoil concerning plastic recycling (Wang et al., 2020), forcing other economies to increase their waste treatment capacity. It was not possible for other large economies to immediately replace the role of China (Huang et al., 2020). This has led to two impacts on Africa. First, the price of used PET bottles fell without an end-market, impacting existing recyclers. Second, high-income countries, along with implementing some wastes reduction strategies (Wang et al., 2020), have shifted their exports to other or low or medium-income countries, which are often ill-equipped to deal with the influx of waste (Brooks et al., 2018). This raises concerns regarding plastic waste imports into Africa. In 2019, to control exports and imports of most plastic scrap and waste, Parties to the Basel Convention adopted plastic scrap and waste amendments. These amendments apply to countries party to the convention (see Chap. 4 for contracting countries) and require prior notice and consent from importing and transit countries before transboundary movement can occur. These amendments took effect on January 1, 2021, and should lead to benefits for African markets with adequate enforcement.

Whether the bulk of plastic used is a domestic product or imported is country-dependent. However, considering imports of polymers and plastics for Africa as a whole, Babayemi et al. (2019) estimated that roughly 172 MT were imported into Africa between 1990 and 2017 (excluding cars, electronics, and sports equipment). Most (51%) were imported into Egypt, Nigeria, South Africa, Algeria, Morocco, and Tunisia (in decreasing order). A further calculation on the end-of-life estimated that between 1990 and 2017, 82.4 MT of plastic waste was produced (in the 33 countries with data, which formed 117.6 MT of the 172 MT). Without policy change, plastic import volumes are forecasted (pre-COVID-19) to double by 2030 (Babayemi et al., 2019; Jambeck et al., 2018). Without a proper waste management plan and strategy, there will be a build-up of billions of tonnes of plastics within both terrestrial and aquatic environments in the near future (Geyer et al., 2017).

It has already been recognised that not all international and high-income country solutions to waste management and marine litter are relevant to an African context (The African Development Bank, 2002). Rather there is a need to focus on and support innovative African-based solutions (UNEP, 2018b), and circular economy initiatives. Taking this into account, the African continent has great opportunities to leapfrog across ineffective systems and approaches which have been applied and now engrained elsewhere. Global emergencies, such as COVID-19, and the inward response of countries and regions, further highlight this need.

Box 1.2: Impact and Uncertainty Driven by COVID-19

COVID-19 has had a profound effect on waste production. This includes increases in personal protective equipment (masks, gloves etc.-both used in health care and privately) (Fadare & Okoffo, 2020; Prata et al., 2020; Zambrano-Monserrate et al., 2020), and an unconfirmed, speculated increase in other single-use items (as people become nervous of reuse in a public pandemic environment), and packaging from increased home delivery (Vanapalli et al., 2020). Africa is particularly susceptible to the increase in personal protective equipment in hospitals as solid medical waste management in most African countries is sub-standard (Udofia et al., 2015). The impact of the COVID-19 pandemic on the existing plastic pollution problem in Africa is highlighted by Benson et al. (2021), who estimate that over 12 billion medical and fabric face masks (105,000 tonnes) are discarded monthly in Africa, with 15 countries considered significant contributors:

Nigeria (15%), Ethiopia (9%), Egypt (8%), DR Congo (7%), Tanzania (5%), and South Africa (4%).

COVID-19 has the potential to have long-term social and economic impacts on Africa.

The negative **economic** impact of COVID-19 may divert vital resources from waste management infrastructure projects and solutions to the rising plastic waste problem and marine litter—both through international investments and regional and local funds. Before the start of the COVID-19 pandemic, Africa was the second fastest-growing region globally, with annual economic growth of 3.4–3.7% (International Monetary Fund, 2021; United Nations, 2020). For the first time in a decade, investment expenditure rather than consumption accounted for more than half the GDP growth (United Nations, 2020). Pre-COVID-19 projections forecast Africa's economy to grow, despite external shocks, including a commodities shock caused by a decrease in demand from China and a resulting recession in commodities impacting African economies between 2018 and 2020 (The Economist, 2021). The economic hardship and recession caused by COVID-19 and the diversion of funds to tackle the pandemic could lead to African governments placing less budget on waste management issues, and hence an increased fraction of wastes could enter the ocean as marine litter.

The pandemic, associated lockdowns, and the downturn in global trade led to an observed decrease of −2.1% real GDP (African Development Bank Group, 2021). Lone and Ahmad (2020) provide a review, summarising the economic damage caused by COVID-19 in Africa—what it has generated so far, and what is still expected to come. Recovery is expected to vary across the region (The Economist, 2021), and contrasting predictions occur. Growth on the continent was forecast to rebound to 5% in 2021, if supported by effective response measures and global economic recovery (United Nations, 2020). However, compared globally, Africa is showing and is predicted to have, a slower recovery from the economic and social effects of COVID-19 compared to more developed regions (The Economist, 2021; United Nations, 2022). Vaccine inequality has contributed (United Nations, 2022) to a slow vaccination rollout in Africa, dragging out the pandemic on the continent (The Economist Intelligence Unit, 2021). An extended pandemic through slow vaccine rollout will have drastic social and economic effects on Africa. Economic output in Africa is projected to rise at a lower and slower rate than pre-pandemic projections. For example, 2023 projections for Africa show a gap of 5.5% compared to pre-pandemic economic growth projection (United Nations, 2022).

There is a danger that the COVID-19 pandemic could push 27 million Africans into extreme poverty, exacerbate existing income inequalities, especially in health and education, and ignite the first recession in Africa in 25 years with a GDP loss of $62.8 billion (United Nations, 2020; Zeufack et al., 2020). Before the pandemic, there was a need for investment in education and infrastructure for good returns in long-term GDP (United Nations, 2020). Socially, in addition to widespread school closures affecting education and household income loss, Africa is likely to see a reduction in female education specifically (The Economist, 2021). Female education is directly linked to proactive family planning, and reduced family size (Subbarao & Raney, 1995), reducing population growth. Any adverse effect on female education through economic and social stresses can further increase population growth in Africa.

Acknowledgements We would like to acknowledge the valuable insights proved by Clever Mafuta, Morten Sorensen, Peter Ryan, Anham Salyani, Salieu Sankoh, and Tony Ribbink in peer-reviewing this chapter. We would also like to acknowledge Nieves López and Federico Labanti (Studio Atlantis) for creating the illustrations.

References

AfDB, UNEP, & GRID-Arendal. (2020). *Sanitation and wastewater atlas of Africa*. Abidjan, Nairobi and Arendal. https://www.grida.no/publications/471

African Development Bank Group. (2021). *African economic outlook 2021: From debt resolution to growth: The road ahead for Africa*. https://www.afdb.org/en/knowledge/publications/african-economic-outlook

Alimi, O. S., Fadare, O. O., & Okoffo, E. D. (2021). Microplastics in African ecosystems: Current knowledge, abundance, associated contaminants, techniques, and research needs. *Science of the Total Environment*. https://doi.org/10.1016/j.scitotenv.2020.142422

APWC (Asia Pacific Waste Consultants). (2020). *Port reception waste facilities audit*. South Africa. https://www.cefas.co.uk/clip/resources/reports/africa-clip-reports/port-reception-waste-facilities-audit-south-africa-apwc/

AU-IBAR. (2019a). *Africa Blue Economy strategy*. https://www.au-ibar.org/sites/default/files/2020-10/sd_20200313_africa_blue_economy_strategy_en.pdf

AU-IBAR. (2019b). *Annex 2. Shipping/transportation, trade, ports, maritime security, safety and enforcement in the context of Africa Blue Economy*. Nairobi, Kenya. http://repository.au-ibar.org/bitstream/handle/123456789/424/Annex%202_AU%20BE%20Shipping%2C%20transportation%2C%20trade%2C%20ports%2C%20maritime%20security%2C%20safety%20and%20enforcement_English.pdf?sequence=1&isAllowed=y

Awumbila, M. (2017). *United Nations expert group meeting on drivers of migration and urbanization in Africa: Key trends and issues*. https://www.un.org/development/desa/pd/sites/www.un.org.development.desa.pd/files/unpd_egm_201709_s3_paper-awunbila-final.pdf

Babayemi, J. O., Nnorom, I. C., Osibanjo, O., & Weber, R. (2019). Ensuring sustainability in plastics use in Africa: Consumption, waste generation, and projections. *Environmental Science Europe, 31*, 60. https://doi.org/10.1186/s12302-019-0254-5

Bakir, A., Rowland, S. J., Thompson, R. C. (2012). Competitive sorption of persistent organic pollutants onto microplastics in the marine environment. *Marine Pollution Bulletin, 64*, 2782–2789. https://doi.org/10.1016/j.marpolbul.2012.09.010

Barboza, L. G. A., Cózar, A., Gimenez, B. C. G., Barros, T. L., Kershaw, P. J., & Guilhermino, L. (2019). Macroplastics pollution in the marine environment. In C. Sheppard (Ed.), *World seas:*

An environmental evaluation (pp. 305–328). Academic Press. https://doi.org/10.1016/B978-0-12-805052-1.00019-X

Belhabib, D., Greer, K., & Pauly, D. (2018). Trends in industrial and artisanal catch per effort in West African fisheries. *Conservation Letters, 11.* https://doi.org/10.1111/conl.12360

Belhabib, D., Sumaila, U. R., Lam, V. W. Y., Zeller, D., Billon, P. L., Kane, E. A., & Pauly, D. (2015). Euros vs. Yuan: Comparing European and Chinese fishing access in West Africa. *PLoS One, 10.* https://doi.org/10.1371/journal.pone.0118351

Benson, N. U., Fred-Ahmadu, O. H., Bassey, D. E., & Atayero, A. A. (2021). COVID-19 pandemic and emerging plastic-based personal protective equipment waste pollution and management in Africa. *Journal of Environmental Chemical Engineering, 9,* 105222. https://doi.org/10.1016/j.jece.2021.105222

Biermann, L., Clewley, D., Martinez-Vicente, V., & Topouzelis, K. (2020). Finding plastic patches in coastal waters using optical satellite data. *Scientific Reports, 10.* https://doi.org/10.1038/s41598-020-62298-z

Bindoff, N. L., Cheung, W. W. L., Kairo, J. G., Aristegui, J., Guinder, V. A., Hallberg, R., Hilmi, N. J. M., Jiao, N., Karim, M. S., Levin, L., O'Donoghue, S., Purca Cuicapusa, S. R., Rinkevich, B., Suga, T., Tagliabue, A., & Williamson, P. (2019). Changing ocean, marine ecosystems, and dependent communities. In H.-O. Portner, D. C. Roberts, V. Masson-Delmotte, P. Zhai, M. Tignor, E. Poloczanska, K. Mintenbeck, A. Alegria, M. Nicolai, A. Okem, J. Petzold, B. Rama, & N. M. Weyer (Eds.), *PCC special report on the ocean and cryosphere in a changing climate* (pp. 447–588). In press. https://www.ipcc.ch/site/assets/uploads/sites/3/2019/11/09_SROCC_Ch05_FINAL-1.pdf

Bonanno, G., & Orlando-Bonaca, M. (2020). Marine plastics: What risks and policies exist for seagrass ecosystems in the Plasticene. *Marine Pollution Bulletin, 158.* https://doi.org/10.1016/j.marpolbul.2020.111425

Borrelle, S. B., Ringma, J., Law, K. L., Monnahan, C. C., Lebreton, L., Mcgivern, A., Murphy, E., Jambeck, J., Leonard, G. H., Hilleary, M. A., & Eriksen, M. (2020). Predicted growth in plastic waste exceeds efforts to mitigate plastic pollution. *Science, 80, 1518,* 1515–1518. https://doi.org/10.1126/science.aba3656

Brooks, A. L., Wang, S., & Jambeck, J. R. (2018). The Chinese import ban and its impact on global plastic waste trade. *Science Advances, 4.* https://doi.org/10.1126/sciadv.aat0131

Browne, M. A., Crump, P., Niven, S. J., Teuten, E., Tonkin, A., Galloway, T., & Thompson, R. (2011). Accumulation of microplastic on shorelines woldwide: Sources and sinks. *Environmental Science & Technology, 45,* 9175–9179. https://doi.org/10.1021/es201811s

Carr, S. A., Liu, J., & Tesoro, A. G. (2016). Transport and fate of microplastic particles in wastewater treatment plants. *Water Research, 91.* https://doi.org/10.1016/j.watres.2016.01.002

Clarke, J. I., Kröner, A., Smedley, A., Nicol, D. S. H. W., Mabogunje, A. L., Middleton, J. F. M., et al. (2020). Africa. *Encyclopedia Britannica.*

Cole, M., Lindeque, P., Fileman, E., Halsband, C., & Galloway, T. S. (2015). The impact of polystyrene microplastics on feeding, function and fecundity in the marine copepod *Calanus helgolandicus. Environmental Science and Technology, 49,* 1130–1137. https://doi.org/10.1021/es504525u

Cole, M., Lindeque, P. K., Fileman, E., Clark, J., Lewis, C., Halsband, C., & Galloway, T. S. (2016). Microplastics alter the properties and sinking rates of zooplankton faecal pellets. *Environmental Science and Technology, 50,* 3239–3246. https://doi.org/10.1021/acs.est.5b05905

Costanza, R., D'Arge, R., de Groot, R., Farber, S., Grasso, M., Hannon, B., et al. (1997). The value of the world's ecosystem services and natural capital. *Nature, 387,* 253–260. https://doi.org/10.1038/387253a0

Costello, C., Ovando, D., Clavelle, T., Strauss, C. K., Hilborn, R., Melnychuk, M. C., et al. (2016). Global fishery prospects under contrasting management regimes. *Proceedings of the National Academy of Sciences of the United States of America, 113,* 5125–5129. https://doi.org/10.1073/pnas.1520420113

Costello, L., McDermott, M. L., Patel, P., Dare, J. (2019). 'A lot better than medicine' - Self-organised ocean swimming groups as facilitators for healthy ageing. *Health & Place, 60.* https://doi.org/10.1016/j.healthplace.2019.102212

De Kock, L., Sadan, Z., Arp, R., & Upadhyaya, P. (2020). A circular economy response to plastic pollution: Current policy landscape and consumer perception. *South African Journal of Science, 116.* https://doi.org/10.17159/sajs.2020/8097

de Smit, J. C., Anton, A., Martin, C., Rossbach, S., Bouma, T. J., & Duarte, C. M. (2021). Habitat-forming species trap microplastics into coastal sediment sinks. *Science of the Total Environment, 772.* https://doi.org/10.1016/j.scitotenv.2021.145520

Deng, H., He, J., Feng, D., Zhao, Y., Sun, W., Yu, H., & Ge, C. (2021). Microplastics pollution in mangrove ecosystems: A critical review of current knowledge and future directions. *Science of the Total Environment, 753,* 142041. https://doi.org/10.1016/j.scitotenv.2020.142041

Diaz, R., Selman, M., Chique, C. (2011). *Global eutrophic and hypoxic coastal systems.* World Resources Institute. Eutrophication and hypoxia: Nutrient pollution in coastal waters.

Diversitas. (2009). What are coral reef services worth? $130,000 to $1.2 million per hectare, per year. *ScienceDaily.* www.sciencedaily.com/releases/2009/10/091016093913.htm

Duarte, C. M. (2017). Reviews and syntheses: Hidden forests, the role of vegetated coastal habitats in the ocean carbon budget. *Biogeosciences, 14,* 301–310. https://doi.org/10.5194/bg-14-301-2017

Duarte, C., Losada, I., Hendriks, I., Mazarras, I., & Marbà, N. (2013). The role of coastal plant communities for climate change mitigation and adaptation. *Nature Climate Change, 3,* 961–968. https://doi.org/10.1038/nclimate1970

Duhec, A. V., Jeanne, R. F., Maximenko, N., & Hafner, J. (2015). Composition and potential origin of marine debris stranded in the Western Indian Ocean on remote Alphonse Island, Seychelles. *Marine Pollution Bulletin, 96,* 76–86. https://doi.org/10.1016/j.marpolbul.2015.05.042

Eriksen, M., Lebreton, L. C. M., Carson, H. S., Thiel, M., Moore, C. J., & Borerro, J. C. (2014). Plastic pollution in the world's oceans: More than 5 trillion plastic pieces weighing over 250,000 tons afloat at sea. *PLoS One.* https://doi.org/10.1371/journal.pone.0111913

Eunomia. (2017). *Investigating options for reducing releases in the aquatic environment of microplastics emitted by (but not intentionally added in) products—Interim report.* Report for DG Env EC. https://www.eunomia.co.uk/reports-tools/investigating-options-for-reducing-rel eases-in-the-aquatic-environment-of-microplastics-emitted-by-products/

Fadare, O. O., & Okoffo, E. D. (2020). Covid-19 face masks: A potential source of microplastic fibers in the environment. *Science of the Total Environment, 737.* https://doi.org/10.1016/j.scitot env.2020.140279

FAO. (2007). *The world mangroves 1980–2005. A thematic study prepared in the framework of the global forest resources assessment 2005.* https://agris.fao.org/agris-search/search.do?recordID= XF2008433305

Fazey, F. M. C., & Ryan, P. G. (2016). Debris size and buoyancy influence the dispersal distance of stranded litter. *Marine Pollution Bulletin, 110,* 371–377. https://doi.org/10.1016/J.MARPOL BUL.2016.06.039

Field, J. G., Hemple, G., & Summerhayes, C. P. (2002). *Oceans 2020: Science trends, and the challenges of sustainability.* Island Press. https://doi.org/10.5670/oceanog.2003.21

Ford, H. V., Jones, N. H., Davies, A. J., Godley, B. J., Jambeck, J. R., Napper, I. E., et al. (2022). The fundamental links between climate change and marine plastic pollution. *Science of the Total Environment, 806,* 150392. https://doi.org/10.1016/j.scitotenv.2021.150392

Friedlingstein, P., Betts, R., Bopp, L., Bloh, W. V., Brovkin, V., & Doney, S. (2006). Climate–carbon cycle feedback analysis, results from the C4MIP model intercomparison. *Journal of Climate, 19,* 3337–3353. https://doi.org/10.1175/JCLI3800.1

Fung, I. Y., Doney, S. C., Lindsay, K., & John, J. (2005). Evolution of carbon sinks in a changing climate. *Proceedings of the National Academy of Sciences of the United States of America, 102,* 11201–11206. https://doi.org/10.1073/pnas.0504949102

GESAMP. (1991). *The state of the marine environment.* http://www.gesamp.org/publications/the-state-of-the-marine-environment

GESAMP. (2015). *Sources, fate and effects of microplastics in the marine environment: A global assessment.* https://ec.europa.eu/environment/marine/good-environmental-status/descriptor-10/ pdf/GESAMP_microplastics%20full%20study.pdf

GESAMP. (2019). *Guidelines for the monitoring and assessment of plastic litter in the ocean.* GESAMP. https://www.researchgate.net/publication/332014608_GESAMP_2019_Guidelines_for_the_monitoring_assessment_of_plastic_litter_in_the_ocean_Reports_Studies_99_editors_Kershaw_PJ_Turra_A_and_Galgani_F

Geyer, R., Jambeck, J. R., & Law, K. L. (2017). Production, use, and fate of all plastics ever made. *Science Advances, 3*, 25–29. https://doi.org/10.1126/sciadv.1700782

Godfrey, L., Ahmed, M. T., Gebremedhin, K. G., Katima, J. H. Y., Oelofse, S., & Osibanjo, O. (2019). *Solid waste management in Africa: Governance failure or development opportunity?* https://doi.org/10.5772/intechopen.86974

Green, E. P., & Short, F. T. (2003). *World atlas of seagrasses.* University of California Press. https://www.researchgate.net/publication/269988511_World_Atlas_of_Seagrasses

Gruber, N., Clement, D., Carter, B. R., Feely, R. A., Van Heuven, S., Hoppema, M., et al. (2019). The oceanic sink for anthropogenic CO_2 from 1994 to 2007. *Science, 80, 363*, 1193–1199. https://doi.org/10.1126/science.aau5153

Halwart, M. (2020). Fish farming high on the global food system agenda in 2020. *FAO Aquaculture Newsletter,* II–III.

Harris, P. T., Westerveld, L., Nyberg, B., Maes, T., Macmillan-Lawler, M., & Appelquist, L. R. (2021). Exposure of coastal environments to river-sourced plastic pollution. *Science of the Total Environment, 769.* https://doi.org/10.1016/j.scitotenv.2021.145222

Harvey, B. P., Kerfahi, D., Jung, Y., Shin, J., Adams, J. M., & Hall-Spencer, J. H. (2020). Ocean acidification alters bacterial communities on marine plastic debris. *Marine Pollution Bulletin, 161*, 111749. https://doi.org/10.1016/j.marpolbul.2020.111749

Henderson, J. V., Storeygard, A., & Deichmann, U. (2017). Has climate change driven urbanization in Africa? *Journal of Development Economics, 124*, 60–82. https://doi.org/10.1016/j.jdeveco.2016.09.001

Hoegh-Guldberg, O., Thezar, M., Boulos, M., Guerraoui, M., Harris, A., Graham, A., et al. (2015). *Reviving the ocean economy: The case for action—2015.* WWF International. https://wwfint.aws assets.panda.org/downloads/reviving_ocean_economy_report_hi_res.pdf

Hoornweg, D., & Bhada-Tata, P. (2012). *What a waste—A global review of solid waste management.* https://openknowledge.worldbank.org/handle/10986/17388

Hoornweg, D., Bhada-Tata, P., & Kennedy, C. (2013). Environment: Waste production must peak this century. *Nature, 502*, 615–617. https://doi.org/10.1038/502615a

Huang, Q., Chen, G., Wang, Y., Chen, S., Xu, L., & Wang, R. (2020). Modelling the global impact of China's ban on plastic waste imports. *Resources, Conservation & Recycling, 154.* https://doi.org/10.1016/j.resconrec.2019.104607

International Monetary Fund. (2019). *GDP.* https://www.imf.org/external/datamapper/NGDPD@WEO/OEMDC/ADVEC/WEOWORLD/AFQ

International Monetary Fund. (2021). *IMF DataMapper: Real GDP growth.* https://www.imf.org/external/datamapper/NGDP_RPCH@WEO/OEMDC/ADVEC/WEOWORLD/AFQ

IPCC. (2019). *Summary for policymakers. Climate change and land: An IPCC special report on climate change, desertification, land degradation, sustainable land management, food security, and greenhouse gas fluxes in terrestrial ecosystems.* In press.

IUCN, EA, & QUANTIS. (2020a). *National guidance for plastic pollution hotspotting and shaping action, country report Kenya.* United Nations Environment Programme. https://plastichotspotting.lifecycleinitiative.org/wp-content/uploads/2020/12/kenya_final_report_2020.pdf

IUCN, EA, & QUANTIS. (2020b). *National guidance for plastic pollution hotspotting and shaping action, country report Mozambique.* United Nations Environment Programme. https://plastichotspotting.lifecycleinitiative.org/wp-content/uploads/2020/12/mozambique_final_report_2020.pdf

IUCN, EA, & QUANTIS. (2021). *National guidance for plastic pollution hotspotting and shaping action, country report Tanzania.* https://www.iucn.org/sites/dev/files/content/documents/tanzania_-_national_guidance_for_plastic_pollution_hotspotting_and_shaping_action_-_2021.pdf

Jambeck, J., Hardestry, B. D., Brooks, A. L., Friend, T., Teleki, K., Fabres, J., et al. (2018). Challenges and emerging solutions to the land-based plastic waste issue in Africa. *Marine Policy, 96*, 256–263. https://doi.org/10.1016/j.marpol.2017.10.041

Jambeck, J. R., Geyer, R., Wilcox, C., Siegler, T. R., Perryman, M., Andrady, A., et al. (2015). Plastic waste inputs from land into the ocean. *Science, 80, 347*, 768–771. https://doi.org/10.1126/science.1260352

Kaiser, M., Attrill, M., Jennings, S., Thomas, D. N., Barnes, D. K., Brierley, A. S., et al. (2005). *Marine ecology: Processes, systems and impacts.* Oxford University Press. https://www.akademika.no/marine-ecology/9780198717850

Karani, P., & Failler, P. (2020). Comparative coastal and marine tourism, climate change, and the blue economy in African Large Marine Ecosystems. *Environmental Development, 36.* https://doi.org/10.1016/j.envdev.2020.100572

Kaza, S., Yao, L. C., Bhada-Tata, P., & Van Woerden, F. (2018). *What a Waste 2.0: A global snapshot of solid waste management to 2050.* Washington, DC. https://openknowledge.worldbank.org/handle/10986/30317

Koelmans, A. A., Bakir, A., Burton, G. A., Janssen, C. R. (2016). Microplastic as a vector for chemicals in the aquatic environment: Critical review and model-supported reinterpretation of empirical studies. *Environmental Science & Technology, 50*, 3315–3326. https://doi.org/10.1021/acs.est.5b06069

Lam, V. W. Y., Cheung, W. W., Swartz, W., & Sumaila, U. R. (2012). Climate change impacts on fisheries in West Africa: Implications for economic, food and nutritional security. *African Journal of Marine Science, 34.* https://doi.org/10.2989/1814232X.2012.673294

Lamb, J. B., Willis, B. L., Fiorenza, E. A., Couch, C. S., Howard, R., Rader, D. N., et al. (2018). Plastic waste associated with disease on coral reefs. *Science, 80, 359*, 460. https://doi.org/10.1126/science.aar3320

Lau, W. W. Y., Shiran, Y., Bailey, R. M., Cook, E., Stuchtey, M. R., Koskella, J., et al. (2020). Evaluating scenarios toward zero plastic pollution. *Science, 80, 369*, 1455–1461. https://doi.org/10.1126/science.aba9475

Lebreton, L., & Andrady, A. (2019). Future scenarios of global plastic waste generation and disposal. *PALGRAVE Communications, 5.* https://doi.org/10.1057/s41599-018-0212-7

Lebreton, L. C. M., Van der Zwet, J., Damsteeg, J. W., Slat, B., Andrady, A., & Reisser, J. (2017). River plastic emissions to the world's oceans. *Nature Communications, 8*, 15611.

Lee, K. W., Shim, W. J., Kwon, O. Y., & Kang, J. H. (2013). Size-dependent effects of micro polystyrene particles in the marine copepod *Tigriopus japonicus. Environmental Science and Technology, 47*, 11278–11283. https://doi.org/10.1021/es401932b

Lewis, S. L., Lopez-Gonzalez, G., Sonké, B., Affum-Baffoe, K., Baker, T. R., Ojo, L. O., et al. (2009). Increasing carbon storage in intact African tropical forests. *Nature, 457*, 1003–1006. https://doi.org/10.1038/nature07771

LME Hub. (2021). *Large marine ecosystems hub.* https://www.lmehub.net/

Lobelle, D., & Cunliffe, M. (2011). Early microbial biofilm formation on marine plastic debris. *Marine Pollution Bulletin, 62.* https://doi.org/10.1016/j.marpolbul.2010.10.013

Lone, S. A., & Ahmad, A. (2020). COVID-19 pandemic—An African perspective. *Emerging Microbes & Infections, 9*, 1300–1308. https://doi.org/10.1080/22221751.2020.1775132

Lutjeharms, J. R. E., & Bornman, T. G. (2010). The importance of the greater Agulhas Current is increasingly being recognized. *South African Journal of Science, 106*, 1–4. https://doi.org/10.4102/sajs.v106i3/4.160

Maes, T., & Preston-Whyte, F. (2022). E-waste it wisely: Lessons from Africa. *SN Applied Sciences, 4*, 72. https://doi.org/10.1007/s42452-022-04962-9

Malone, T. C., & Newton, A. (2020). The globalization of cultural eutrophication in the coastal ocean: Causes and consequences. *Frontiers in Marine Science, 7.* https://doi.org/10.3389/fmars.2020.00670

Martin, C., Baalkhuyur, F., Valluzzi, L., Saderne, V., Cusack, M., & Almahasheer, H. (2020). Exponential increase of plastic burial in mangrove sediments as a major plastic sink. *Science Advances, 6*, 5593. https://doi.org/10.1126/sciadv.aaz5593

Masnadi, M. S., El-Houjeiri, H. M., Schunack, D., Li, Y., Englander, J. G., Badahdah, A., et al. (2018). Global carbon intensity of crude oil production. *Science, 80, 361*, 851–853. https://doi.org/10.1126/science.aar6859

Mcleod, E., Chmura, G. L., Bouillon, S., Salm, R., Björk, M., Duarte, C. M., et al. (2011). A blueprint for blue carbon: Toward an improved understanding of the role of vegetated coastal habitats in sequestering CO_2. *Frontiers in Ecology and the Environment, 9*, 552–560. https://doi.org/10.1890/110004

Menicagli, V., Balestri, E., Vallerini, F., De Battisti, D., & Lardicci, C. (2021). Plastics and sedimentation foster the spread of a non-native macroalga in seagrass meadows. *Science of the Total Environment, 757*. https://doi.org/10.1016/j.scitotenv.2020.143812

Moss, K., Allen, D., González-Fernández, D., & Allen, S. (2021). Filling in the knowledge gap: Observing MacroPlastic litter in South Africa's rivers. *Marine Pollution Bulletin, 162*. https://doi.org/10.1016/j.marpolbul.2020.111876

Näkki, P., Eronen-Rasimus, E., Kaartokallio, H., Kankaanpää, H., Setälä, O., Vahtera, E., Lehtiniemi, M. (2021). Polycyclic aromatic hydrocarbon sorption and bacterial community composition of biodegradable and conventional plastics incubated in coastal sediments. *Science of the Total Environment, 755*. https://doi.org/10.1016/j.scitotenv.2020.143088

Napper, I. E., & Thompson, R. C. (2016). Release of synthetic microplastic plastic fibres from domestic washing machines: Effects of fabric type and washing conditions. *Marine Pollution Bulletin, 112*, 39–45. https://doi.org/10.1016/j.marpolbul.2016.09.025

Ncube, M., & Lufumpa, C. L. (2014). *The emerging middle class in Africa*. Routledge. https://www.routledge.com/The-Emerging-Middle-Class-in-Africa/Ncube-Lufumpa/p/book/9781138796430

Niang, I., Ruppel, O. C., Abdrabo, M. A., Essel, A., Lennard, C., Padgham, J., et al. (2014). Africa. In *Barros, impacts, adaptation, and vulnerability. Part B: Regional aspects* (pp. 1199–1265). Contribution of working group II to the fifth assessment report of the intergovernmental panel on climate change. Cambridge University Press. https://www.researchgate.net/publication/309475977_Chapter_22_Africa_In_Climate_Change_2014_Impacts_Adaptation_and_Vulnerability_Part_B_Regional_Aspects_Contribution_of_Working_Group_II_to_the_Fifth_Assessment_Report_of_the_Intergovernmental_Panel_on_

OECD. (2016). *The ocean economy in 2030*. https://doi.org/10.1787/9789264251724-en

Oppenheimer, M., Glavovic, B. C., Hinkel, J., van de Wal, R., Magnan, A. K., Abd-Elgawad, A., et al. (2019). Sea level rise and implications for low-lying islands, coasts and communities. In *IPCC special report on the ocean and cryosphere in a changing climate* (pp. 321–446). In press. https://www.ipcc.ch/site/assets/uploads/sites/3/2019/11/08_SROCC_Ch04_FINAL.pdf

Prata, J. C., Silva, A. L. P., Walker, T. R., Duarte, A. C., & Rocha-santos, T. (2020). COVID-19 pandemic repercussions on the use and management of plastics. *Environmental Science & Technology, 54*, 7760–7765. https://doi.org/10.1021/acs.est.0c02178

Reichert, J., Schellenberg, J., Schubert, P., & Wilke, T. (2018). Responses of reef building corals to microplastic exposure. *Environmental Pollution, 237*, 955–960. https://doi.org/10.1016/j.envpol.2017.11.006

Rochman, C. M., Brookson, C., Bikker, J., Djuric, N., Earn, A., Bucci, K., Athey, S., Huntington, A., McIlwraith, H., Munno, K., Frond, H. De, Kolomijeca, A., Erdle, L., Grbic, J., Bayoumi, M., Borrelle, S. B., Wu, T., Santoro, S., Werbowski, L. M., Zhu, X., Giles, R. K., Hamilton, B. M., Thaysen, C., Kaura, A., Klasios, N., Ead, L., Kim, J., Sherlock, C., Ho, A., Hung, C. (2019). Rethinking microplastics as a diverse contaminant suite. *Environmental Toxicology and Chemistry, 38*, 703–711. https://doi.org/10.1002/etc.4371

Ryan, P. G. (2020). The transport and fate of marine plastics in South Africa and adjacent oceans. *South African Journal of Science, 116*, 5–6. https://doi.org/10.17159/sajs.2020/7677

Ryan, P. G., Dilley, B. J., Ronconi, R. A., & Connan, M. (2019). Rapid increase in Asian bottles in the South Atlantic Ocean indicates major debris inputs from ships. *Proceedings of the National Academy of Sciences of the United States of America, 116*, 20892–20897. https://doi.org/10.1073/pnas.1909816116

Ryan, P. G., & Perold, V. (2021). Limited dispersal of riverine litter onto nearby beaches during rainfall events. *Estuarine, Coastal and Shelf Science, 238*, 1008–1016. https://scholar.google.com/citations?view_op=view_citation&hl=en&user=jZbtaTwAAAAJ&citation_for_view=jZbtaTwAAAAJ:3fE2CSJIrl8C

Ryan, P. G., Perold, V., Osborne, A., & Moloney, C. L. (2018). Consistent patterns of debris on South African beaches indicate that industrial pellets and other mesoplastic items mostly derive from local sources. *Environmental Pollution.* https://doi.org/10.1016/j.envpol.2018.02.017

Ryan, P. G., Weideman, E. A., Perold, V., Hofmeyr, G., & Connan, M. (2021). Message in a bottle: Assessing the sources and origins of beach litter to tackle marine pollution. *Enviornmental Justice, 288.* https://doi.org/10.1016/j.envpol.2021.117729

Sabine, C. L., Heimann, M., Artaxo, P., Bakker, D., Chen, C. T., Field, C. B., et al. (2004). Current status and past trend of the global carbon cycle. In C. B. Field & M. R. Raupach (Eds.), *The global carbon cycle: Integrating humans, climate, and the natural world, scope 62* (pp. 17–44). Island Press. https://globalchange.mit.edu/publication/13838

Sala, E., Lubchenco, J., Grorud-Colvert, K., Novelli, C., Roberts, C., & Sumaila, U. R. (2018). Assessing real progress towards effective ocean protection. *Marine Policy.* https://doi.org/10.1016/j.marpol.2018.02.004

Satia, B. P. (2016). An overview of the large marine ecosystem programs at work in Africa today. *Environment and Behaviour, 17,* 11–19. https://doi.org/10.1016/j.envdev.2015.06.007

Scharrer, T., O'Kane, D., & Kroeker, L. (2018). Introduction: Africa's middle classes in critical perspective. In *Middle classes in Africa: Changing lives and conceptual challenges* (pp. 1–31). Springer. https://doi.org/10.1007/978-3-319-62148-7_1

Seng, N., Lai, S., Fong, J., Saleh, M. F., Cheng, C., Cheok, Z. Y., et al. (2020). Early evidence of microplastics on seagrass and macroalgae. *Marine and Freshwater Research, 71.* https://doi.org/10.1071/MF19177

Severin, M. I., Vandegehuchte, M. B., Hooyberg, A., Buysse, A., Raes, F., Everaert, G. (2021). Influence of the belgian coast on well-being during the COVID-19 pandemic. *Psychologica Belgica, 61,* 284–295. https://doi.org/10.5334/PB.1050

Sheavly, S. B. (2005). *Marine debris: An overview of a critical issue for our oceans.* https://www.scirp.org/(S(351jmbntvnsjt1aadkposzje))/reference/ReferencesPapers.aspx?ReferenceID=1752350

Shen, M., Huang, W., Chen, M., Song, B., Zeng, G., & Zhang, Y. (2020). (Micro)plastic crisis: Un-ignorable contribution to global greenhouse gas emissions and climate change. *Journal of Cleaner Production, 254.* https://doi.org/10.1016/j.jclepro.2020.120138

Sherman, K., O'Reilly, J., Belkin, I. M., Melrose, C., & Friedland, K. D. (2011). The application of satellite remote sensing for assessing productivity in relation to fisheries yields of the world's large marine ecosystems. *ICES Journal of Marine Science, 68,* 667–676. https://doi.org/10.1093/icesjms/fsq177

Subbarao, K., & Raney, L. (1995). Social gains from female education: A cross-national study. *Economic Development and Cultural Change, 44.*

Suyadi, Manullang, C. Y. (2020). Distribution of plastic debris pollution and it is implications on mangrove vegetation. *Marine Pollution Bulletin, 160.* https://doi.org/10.1016/j.marpolbul.2020.111642

Tekman, M. B., Gutow, L., Macario, A., Haas, A., Walter, A., & Bergmann, M. (2019). *Alfred-Wegener-Institut Helmholtz-Zentrum für Polar- und Meeresforschung.* https://litterbase.awi.de/litter_detail

Temmerman, S., Meire, P., Bouma, T. J., Herman, P. M. J., Ysebaert, T., & Vriend, H. J. (2013). Ecosystem-based coastal defence in the face of global change. *Nature, 504,* 79–83. https://doi.org/10.1038/nature12859

The African Development Bank. (2002). *Study on solid waste management options for Africa.* https://www.gcca.eu/sites/default/files/2019-11/2002%20African%20Development%20Bank%20_%20Solid%20Waste%20Management%20Options%20for%20Africa.pdf

The Economist. (2021). *Africa's recovery from covid-19 will be slow.* https://www.economist.com/middle-east-and-africa/2021/02/06/africas-recovery-from-covid-19-will-be-slow

The Economist Intelligence Unit. (2021). *Coronavirus vaccines: Expect delays. Q1 global forecast 2021.* https://www.eiu.com/n/campaigns/q1-global-forecast-2021/

Tramoy, R., Gasperi, J., Colasse, L., & Tassin, B. (2020). Transfer dynamic of macroplastics in estuaries—New insights from the Seine estuary: Part 1. Long term dynamic based on date-prints on stranded debris. *Marine Pollution Bulletin, 152.*

Udofia, E. A., Fobil, J. N., & Gulis, G. (2015). Solid medical waste management in Africa. *African Journal of Environmental Science and Technology, 9*, 244–254. https://doi.org/10.5897/AJEST2 014.1851

UNDESA. (2017). *World population prospects 2017: Data booklet.* https://www.un.org/develo pment/desa/publications/world-population-prospects-the-2017-revision.html

UNDESA. (2019). *World population prospects 2019.* https://reliefweb.int/sites/reliefweb.int/files/resources/WPP2019_Highlights.pdf

UNDESA. (2020). *2020 international trade statistics yearbook. Trade by product. ST/ESA/STAT/SER.G/69* (Vol. II). ISBN 978-92-1-259192-6. https://comtrade.un.org/pb/downloads/2020/VolII2020.pdf

UNEP. (2006). *Marine litter: A global challenge.* UNEP. https://stg-wedocs.unep.org/bitstream/han dle/20.500.11822/31632/MLAGC.pdf?sequence=1&isAllowed=y

UNEP. (2007). *Mangroves of Western and Central Africa.* https://www.unep-wcmc.org/resources-and-data/mangroves-of-western-and-central-africa

UNEP. (2018a). *Africa waste management outlook.* https://wedocs.unep.org/handle/20.500.11822/25514

UNEP. (2018b). *Mapping of global plastics value chain and plastics losses to the environment (with a particular focus on marine environment).* United Nations Environment Programme. https://gefmarineplastics.org/files/2018%20Mapping%20of%20global%20plastics%20value%20chain%20and%20hotspots%20-%20final%20version%20r181023.pdf

UNEP. (2020). *Monitoring plastics in rivers and lakes: Guidelines for the harmonization of methodologies.* https://wedocs.unep.org/bitstream/handle/20.500.11822/35405/MPRL.pdf

UNEP. (2021). *From pollution to solution. A global assessment of marine litter and plastic pollution.* New Scientist. https://doi.org/10.1016/S0262-4079(18)30486-X

United Nations. (2019). *Diving into the blue economy.* https://www.un.org/development/desa/en/news/sustainable/blue-economy.html

United Nations. (2020). *Economic report on Africa 2020: Innovative finance for private sector development in Africa development in Africa.* https://repository.uneca.org/handle/10855/43834

United Nations. (2021). *United Nations goal 14: Conserve and sustainably use the oceans, seas and marine resources, facts & figures.* https://www.un.org/sustainabledevelopment/oceans/.

United Nations. (2022). *World economic situation and prospects 2022.* https://www.un.org/develo pment/desa/dpad/wp-content/uploads/sites/45/publication/WESP2022_web.pdf

UN-HABITAT. (2008). *State of the world's cities 2008/2009, harmonious cities.* London: Earthscan.

Vanapalli, K. R., Sharma, H. B., Ranjan, V. P., Samal, B., Bhattacharya, J., Dubey, B. K., et al. (2020). Challenges and strategies for effective plastic waste management during and post COVID-19 pandemic. *Science of the Total Environment, 750.* https://doi.org/10.1016/j.scitotenv.2020. 141514

Velis, C. A., Lerpiniere, D., & Tsakona, M. (2017). *PREVENT MARINE PLASTIC LITTER—NOW!* International Solid Waste Association. https://marinelitter.iswa.org/fileadmin/user_upload/Mar ine_Task_Force_Report_2017/ISWA_report_interactive.pdf

Vester, C., & Bouwman, H. (2020). Landbased sources and pathways of marine plastics in a South African context. *South African Journal of Science, 116.* https://doi.org/10.17159/sajs.2020/7700

Wang, C., Zhao, L., Lim, M. K., Chen, W. Q., & Sutherland, J. W. (2020). Structure of the global plastic waste trade network and the impact of China's import ban. *Resources, Conservation & Recycling, 153.* https://doi.org/10.1016/j.resconrec.2019.104591

Weideman, E. A., Perold, V., & Ryan, P. G. (2020). Limited long-distance transport of plastic pollution by the Orange-Vaal River system, South Africa. *Science of the Total Environment.* https://doi.org/10.1016/j.scitotenv.2020.138653

Wieczorek, A. M., Croot, P. L., Lombard, F., Sheahan, J. N., & Doyle, T. K. (2019). Microplastic ingestion by gelatinous zooplankton may lower efficiency of the biological pump. *Environmental Science and Technology, 53*, 5387–5395. https://doi.org/10.1021/acs.est.8b07174

Wilson, D., & Velis, C. A. (2015). Waste management—Still a global challenge in the 21st century: An evidence-based call for action. *Waste Management & Research, 33*, 1049–1051. https://doi. org/10.1177/0734242X15616055

Wilson, D. C., Rodic, L., Modak, P., Soos, R., Carpintero, A., Velis, K., et al. (2015). *Global waste management outlook*. International Solid Waste Association and United National Environment Programme. https://www.uncclearn.org/wp-content/uploads/library/unep23092015.pdf

World Bank. (2018). *What a waste global database. Country level dataset*. https://datacatalog.worldbank.org/search/dataset/0039597/What-a-Waste-Global-Database

World Bank. (2019). *DataBank: World development indicators*. https://databank.worldbank.org/reports.aspx?source=2&series=SP.POP.TOTL&country=WLD

World Economic Forum. (2020). *World oceans day: Visualizing our impact on our ocean economy*. Retrieved July 29, 2021, from https://www.weforum.org/agenda/2020/06/human-impact-ocean-economy

World Risk Report. (2018). *Focus child protection and children's rights*. https://reliefweb.int/report/world/world-risk-report-2018-focus-child-protection-and-childrens-rights

Yao, W., Di, D., Wang, Z., Liao, Z., Huang, H., Mei, K., et al. (2019). Micro- and macroplastic accumulation in a newly formed *Spartina alterniflora* colonized estuarine saltmarsh in southeast China. *Marine Pollution Bulletin, 149*, 110636. https://doi.org/10.1016/j.marpolbul.2019.110636

Yasin, J. A., Kroeze, C., Mayorga, E. (2010). Nutrients export by rivers to the coastal waters of Africa: Past and future trends. *Global Biogeochemical Cycles, 24*. https://doi.org/10.1029/2009GB003568

Zambrano-Monserrate, M. A., Ruano, M. A., & Sanchez-Alcalde, L. (2020). Indirect effects of COVID-19 on the environment. *Science of the Total Environment, 728*. https://doi.org/10.1016/j.scitotenv.2020.138813

Zeufack, A. G., Calderon, C., Kambou, G., Kubota, M., Cantu Canales, C., & Korman, V. (2020). *Africa's pulse* (p. 22). https://doi.org/10.1596/978-1-4648-1568-3

Chapter 2
Marine Litter Sources and Distribution Pathways

Takunda Yeukai Chitaka, Percy Chuks Onianwa, and Holly Astrid Nel

Summary Marine litter has been a global concern for many decades. It is important to understand marine litter sources and distribution pathways for the development of targeted and effective interventions and strategies. These have been relatively less researched on the African continent. This chapter focuses on (1) the sources of litter items from macro to nanoscale entering the marine environment and (2) the distribution and accumulation of these items within the environment, focusing on the African marine setting. Case studies are used to showcase specific examples and highlight knowledge/data gaps that need to be addressed within Africa. The potential pathways going forward are discussed and what may be expected in the future, in light of the challenges and successes examined.

Keywords Marine litter · Plastic pollution · Microplastics pollution · Monitoring

2.1 Introduction

Whilst marine litter has been a global concern for many decades. It has been relatively less researched on the African continent (Akindele and Alimba, 2021; Alimi et al., 2021). The majority of quantification studies took place in South Africa, dating back to the 1980s (Ryan, 1988). However, the global spotlight on this issue has seen more studies being conducted across the continent (Fig. 2.1a, b).

T. Y. Chitaka
Department of Social Work, University of the Western Cape, Cape Town, South Africa

P. C. Onianwa
Basel Convention Coordinating Centre for the African Region (in Nigeria), University of Ibadan, Ibadan, Nigeria

Department of Chemistry, University of Ibadan, Ibadan, Nigeria

H. A. Nel (✉)
School of Geography, Earth and Environmental Sciences, University of Birmingham, Birmingham, UK
e-mail: hollynel1988@gmail.com

© The Author(s) 2023
T. Maes and F. Preston-Whyte (eds.), *The African Marine Litter Outlook*,
https://doi.org/10.1007/978-3-031-08626-7_2

35

2.2   Sources of Marine Litter

Marine and freshwater litter (e.g., plastics, ceramics, cloth, glass, metal, paper, rubber, wood) are evident throughout Africa (Chitaka & von Blottnitz, 2019, 2021; Dunlop et al., 2020; Ebere et al., 2019; Moss et al., 2021; Weideman et al., 2020a). The sheer volume littering coasts or floating down rivers highlights its prevalence and the predominance of and leakage from various sources in the region. Sources and release pathways can be linked to land-based or sea-based activities, with the former including municipal solid waste management, direct littering, wastewater and sludge release, agricultural activities, industrial production, harbour/port activities, and others (Fig. 2.2). Sea-based activities include the fishing industry and aquaculture sector and sea-based dumping from ships and off-shore platforms (Fig. 2.2).

2.2.1   Land-Based Sources

Municipal Solid Waste Management and Direct Littering

A major source of litter entering the environment in Africa results from the lack of adequate and appropriate solid waste management, which pervades every country of the continent (UNEP, 2018b). Municipal solid waste generation rates vary across Africa (Hoornweg & Bhada-Tata, 2012; Kaza et al., 2018), however, overall daily capita rates were considered to be 0.78 kg in 2012, compared to a global average of 1.2 kg per capita per day (UNEP, 2018b). Higher waste generation rates have been associated with some African island states (i.e., Seychelles, Mauritius, and Cabo Verde), which have been attributed to the tourism industry and a reliance on imported resources and associated packaging (Andriamahefazafy & Failler, 2021; Hoornweg & Bhada-Tata, 2012).

Increases in waste generation rates are driven by factors that include rapid population growth and urbanisation, a growing middle class with associated changing consumption habits, economic development, and global trade, which encourages imports of consumer goods into Africa (Jambeck et al., 2018). See Chap. 1 for further details on projections for Africa. Despite these projections, service delivery remains poor and is unlikely to improve at rates needed to support the populace. In Sub-Saharan Africa, for example, only about 44% collection rate of waste, on average, is achieved (Kaza et al., 2018). Waste collection and disposal methods are primarily crude. It is estimated that the treatment processes across the region are: open dumping and/or burning (69%), unspecified landfilling (12%), sanitary landfilling (11%), controlled landfilling (1%), and recycling (7%) (Kaza et al., 2018). 19 of the world's 50 biggest dumpsites are in Africa, with six located in Nigeria (UNEP, 2018b). Corresponding data for the North African sub-region alone are not readily available as the area is often combined with that for the

Middle East. In this regard, the Middle East and North Africa region had an estimated average waste generation rate of 0.81 kg per capita per day, which amounted to 129 million tonnes in 2016 (Kaza et al., 2018). The average collection rate was 82% for this combined region but varied significantly amongst the countries. Waste treatment in the region was estimated to be: open dumping (52%), unspecified landfilling (10%), sanitary landfilling (11%), controlled landfilling (14%), recycling (9%), and composting (4%) (Kaza et al., 2018). Waste management data from African Small Island Developing States (SIDS) is limited. However, it is well understood that due to lack of space and infrastructure and disposal sites near the marine environment, SIDS are often disproportionately affected by waste leakage into the environment (see Chap. 3 for more detail). This is often compounded by debris littering beaches brought by ocean currents and higher generation of waste by visiting tourists (UNEP, 2019).

Open spaces where solid waste has been dumped indiscriminately result in high leakage of hazardous and non-hazardous waste into drains and finally into rivers, lakes, and estuaries. The closer the source of mismanaged waste to river networks and coastal zones, the greater the chances of marine litter. Many populated inland cities are located on rivers' banks, which form a rich network of waterways that criss-cross the continent (Grid-Arendal, 2005; Lane et al., 2007; UNEP, 1999). Thus, Africa's inland rivers and estuaries may provide a pathway for a portion of land-derived litter to enter the sea (Lane et al., 2007; Moss et al., 2021; Naidoo & Glassom, 2019; Weideman et al., 2020c) (Fig. 1.1a). Africa has many densely populated coastal cities, of which several have been linked to large litter inputs to the marine environment (Ryan, 2020a, 2020b; Ryan et al., 2021). Lack of adequate affordable housing in Africa may be, for example, a source of litter entering nearby environments, especially as waste is often used as part of informal and temporary structures and shelters (GESAMP, 2019). The coupling between mismanaged waste and affordable housing needs to be investigated further, along with more work on the role coastal cities across Africa play as a source of litter to the surrounding marine environment.

Direct littering and dumping by households in parts of Africa, has also resulted in solid waste entering open drains, river watercourses, and coastal beaches. Beaches in most parts of the world, but especially those in many low-income countries, have been littered with waste by tourists and local persons involved in recreational activities (Lamprecht, 2013; Lane et al., 2007; Tsagbey et al., 2009; UNEP, 2019). Common items include drink bottles, water sachets, single-use food packaging, cigarette butts, and an array of miscellaneous materials. Many beaches in different parts of Africa have been recorded as frequently littered by tourists, especially during peak holiday months (Tsagbey et al., 2009) or specific sporting/entertainment events (Ahmed et al., 2008).

Once in the environment, larger plastic litter items are physically, biologically, and chemically broken down and degraded into secondary fragments/films/foams that include meso, micro, and nano sizes (Bond et al., 2018; Cooper & Corcoran, 2010). The most common polymers detected in African microplastic studies were polyethylene (PE) and polypropylene (PP) (Alimi et al., 2021; Mayoma et al., 2020;

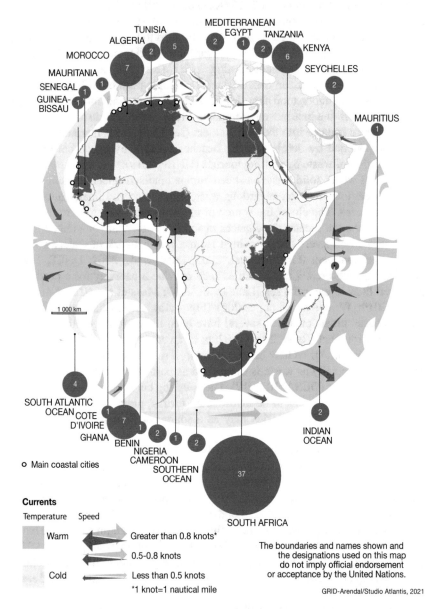

Fig. 2.1 a Total number of marine litter quantification studies published across Africa in peer-reviewed journals (excluding ingestion and entanglement studies, which are covered in Chap. 3). *As of December 2021, detailed list in Annex 2.1. **b** Total number of marine litter quantification studies, by size fractionate, published across Africa in peer-reviewed journals (excluding ingestion and entanglement studies, which are covered in Chap. 3). *As of December 2021, detailed list in Annex 2.1

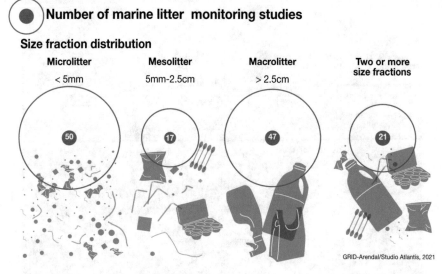

Number of marine litter monitoring studies

Size fraction distribution

Microlitter	Mesolitter	Macrolitter	Two or more size fractions
< 5mm	5mm-2.5cm	> 2.5cm	
50	17	47	21

GRID-Arendal/Studio Atlantis, 2021

Fig. 2.1 (continued)

Missawi et al., 2020; Vetrimurugan et al., 2020; Wakkaf et al., 2020; Zayen et al., 2020), which are widely used in the packaging sector (PlasticsSA, 2018). More targeted studies are needed to investigate the role open or unmanaged dumpsites (Bundhoo, 2018; Nel et al., 2021) and incineration sites, where litter is managed through informal burning (Yang et al., 2021), play in microplastic generation and release. Especially, as these sites may become significant legacy sources, leaching microplastics to the surrounding environment long after site closure.

Wastewater and Sludge

Domestic and industrial wastewater is a well-recognised source of litter that may get deposited in marine environments (Conley et al., 2019; Freeman et al., 2020; GESAMP, 1991; Kay et al., 2018; Okoffo et al., 2019). Domestic and industrial wastewater serve as conduits for litter, which has been purposely dumped/flushed or originated from added products. Once discharged into streams and rivers (or in some countries directly into the marine environment), litter may be carried into the marine environment. In high-income countries with efficient processing plants, the impact of wastewater discharge on the marine environment is usually mitigated by pre-treatment purification steps (biological, chemical, and mechanical). However, whilst such wastewater treatment plants (WWTPs) may remove most macrolitter and a relatively large portion of microlitter, the smaller (<100 μm) litter fractions remain in the effluent, subsequently entering aquatic environments through discharge (Conley et al., 2019; Iyare et al., 2020; Talvitie et al., 2017). Additionally, although microlitter may get removed before the effluent is discharged into aquatic

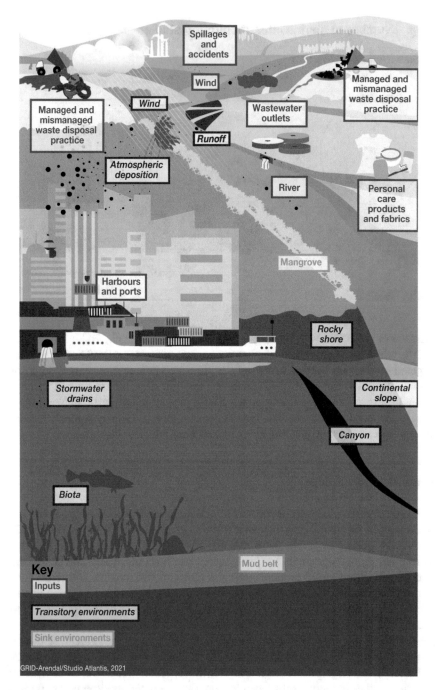

Fig. 2.2 Sources, pathways, and sinks of marine litter from macro- to micro-sized items

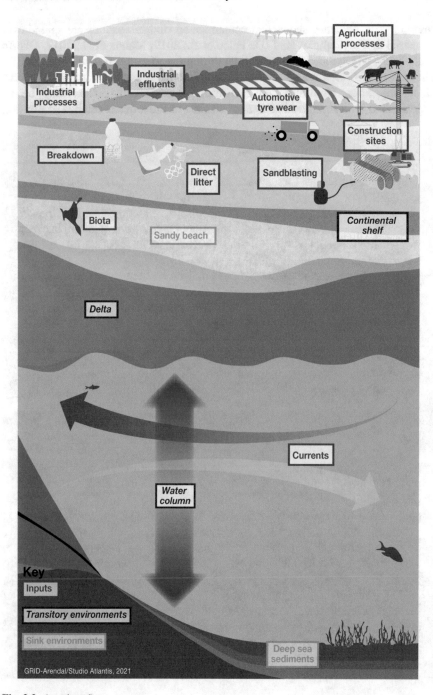

Fig. 2.2 (continued)

environments, evidence shows that contaminated sludge/biosolids are spread over terrestrial and agricultural land (De Falco et al., 2019; Mahon et al., 2017; Okoffo et al., 2020) which then leach microlitter during runoff events (Okoffo et al., 2019). The release of raw/untreated sewage directly into the ocean is an additional source of marine litter. Domestic and industrial wastewater in Africa may be an understudied yet a significant source for all sizes of litter, this is detailed in Box 2.1.

Box 2.1: Domestic and Industrial Wastewater as an Understudied Source for Litter from Macro to Nano

Domestic and industrial wastewater are potential sources of marine litter. An investigation of macrolitter flows in three WWTPs in Cape Town, South Africa, found cotton bud sticks in the discharged effluent, which passed through the primary screens designed to trap debris (Chitaka, 2020). Investigations conducted by Nel et al. (2018) and Dalu et al. (2021) of river sediments upstream and downstream of WWTP effluent discharges in South Africa suggested that WWTPs are point-sources for microplastics and microfibres. Research in the Bizerte Lagoon, Tunisia, and the adjacent coastline identified microplastic hotspots potentially linked to wastewater facilities (Wakkaf et al., 2020).

In most African countries, wastewater management facilities are non-existent (AfDB/UNEP/Grid-Arendal, 2020). Where they exist, they are mostly aging, inadequate, dis-used, or derelict due to high maintenance and replacement costs (Nikiema et al., 2013). Inadequate provisions for sanitation are a significant problem for most African urban, peri-urban, and rural communities. In most African countries, 72–92% of wastewater was untreated in 2015 (WWAP, 2017). Domestic wastewater from households, hospitals, academic institutions, government offices, etc., is rarely discharged into sewers. Instead, most households discharge directly into soak-away pits that contaminate groundwater or open land and public drainage gutters, contaminating rivers, streams, and ultimately marine environments (AfDB/UNEP/Grid-Arendal, 2020; Mafuta et al., 2018; Wang et al., 2014). In some coastal cities, sewage is directly discharged into the sea. SIDS are no exception. Although many households are provided with a supply of water, a wastewater collection/connection is far less common within SIDS (UNEP, 2019). More research needs to be conducted across Africa to assess how the lack of services and poor maintenance of wastewater infrastructure results in the release of all sizes of litter to the natural environment. Thus, allowing us to inform which mitigation methods to prioritise and importantly, which solutions have been effective (Image 2.1).

Image 2.1 Evidence of debris in the **a** dissolved air flotation tank and **b** effluent of a Cape Town WWTP (Chitaka, 2020)

Domestic and industrial wastewater may contain primary microplastics, intentionally incorporated into some products. For example, 'microbeads' made from polyurethane (PU) spheres or PE particles are found in a range of personal care and cosmetic products (Amec Foster Wheeler Environment & Infrastructure UK Limited, 2017; UNEP, 2015). This type of contaminant has gained attention across the globe as a result of the 'Beat the Microbead' campaign (Dauvergne, 2018), which highlighted how these products can get released into the environment via wastewater systems. Microplastics may also be intentionally added to or used in the production of paints/coatings, detergents, slow- and controlled-release fertilisers, and industrial abrasives, and depending on the product, will get released into the environment via wastewater, leaching, and/or stormwater runoff (Amec Foster Wheeler Environment & Infrastructure UK Limited, 2017). Although very high concentrations of microbeads have been recorded within some aquatic environments, for example in sediment from the St. Lawrence River, Canada (10^3 microbeads L^{-1}; Castaneda et al., 2014) and throughout the Irwell and Mersey catchments in the United Kingdom (<70,000 microbeads kg^{-1}; Hurley et al., 2018), similar hotspots have not been detected in Africa. Microbeads associated with personal care products range in size, shape, and colour (Cheung & Fok, 2017). White granule-like PE fragments in face washes may be more challenging to detect than brightly coloured (blue and green) spherical beads found in other cosmetic products (Nel et al., 2019). This difference in detection may result in some microbead granules being overlooked. Regardless, the potential for contamination is apparent, resulting in several countries banning their use in rinse-off products (Guerranti et al., 2019). However, no African country has banned the inclusion of plastic particles in cosmetic products, though discussions have occurred, and some industries (such as the South African cosmetics industry) have implemented some voluntary initiatives to replace microbeads with other materials (Verster & Bouwman, 2020).

Synthetic and natural microfibres can enter the environment as a result of industrial activities (textile factories), individual consumer activities (washing of clothes by hand or using household/communal machines), and wastewater management (sewage effluent, sewage sludge) (Mishra et al., 2019). Marine microplastic studies in South Africa (De Villiers, 2018, 2019; Nel & Froneman, 2015; Nel et al., 2017) and Tunisia (Wakkaf et al., 2020) found microfibres were the most dominant type of microlitter detected. However, by mass of debris they account for <0.01% (Ryan et al., 2020d). Researchers have suggested that microfibres should be classed as their own contaminant independent of 'microplastics'. Anthropogenic fibres recorded in the environment can be plastic in origin. Washing synthetic textile materials has been shown to release large amounts of microfibres into wastewater (Browne et al., 2020; De Falco et al., 2019, 2020). Microfibres can also be natural/non-plastic in origin e.g., cotton, viscose, linen, jute, kenaf, hemp (Suaria et al., 2020a) or synthetic in origin but made from natural sources of regenerated cellulose e.g., rayon (Kanhai et al., 2017). Reports have suggested microplastic and microfibre removal rates of ~90% for treated wastewater, however the combined global release of microlitter annually from untreated effluent has been estimated at 3.85×10^{16} (Pedrotti et al., 2021; Uddin et al., 2020). The largely untreated domestic wastewater that pollutes local streams and rivers in Africa may be expected to be loaded with an abundance of these particles.

Microplastic pollution is expected in industrial wastewaters/drainage, either through intentional industrial processes or as accidental leakage from industries manufacturing items that utilise primary microplastics or process pristine/recycled plastic products (Karlsson et al., 2018). Unfortunately, data on microplastic/microfibre abundance and distribution in industrial wastewaters/drainage is scarce in Africa and globally. However, Zhou et al. (2020b) and Chan et al. (2021) both recorded high levels of pollution (~300/500 microfibres L^{-1}) originating from textile processing factories in China.

Pre-production pellets are another primary microplastic. They have been recorded on beaches in Africa since the 1980s (International Pellet Watch, 2021; Ryan & Moloney, 1990). They have been attributed to unintentional factory and transportation leakage (Boucher & Friot, 2017; Karlsson et al., 2018). Pellets in the marine environment have often been associated with urbanisation and industrialisation centres (Hosoda et al., 2014; Naidoo et al., 2015; Ryan & Moloney, 1990; Ryan et al., 2018). However high concentrations have also been located in more rural locations due to historical deposits resulting from long-range transport carrying high densities of pellets from urban centres (Ryan et al., 2018). For example, in South Africa, pellet deposits are seen at 16 mile Beach in the West Coast National Park and Woody Cape at the east end of Algoa Bay (Ryan et al., 2012, 2018). To combat this specific type of contamination the 'Operation Clean Sweep' campaign has been adopted by plastic producers and converters worldwide (American Chemistry Council, 2021). In Africa, the Egyptian Plastic Exporters and Manufacturers Association, Ghana Plastics Manufacturers Association and Plastics SA have pledged to follow best practice guidelines outlined by the campaign to minimise pellet, flake and powder loss from the plastic industry.

Harbour and Port Activities

Globally, shipping activities are associated with generating large quantities of wastes onboard. At the same time, ports and harbour activities may also generate wastes close to the sea (APWC, 2020; IMO, 1973/1978; Mobilik et al., 2016). Wastes from both sources can contribute to the marine litter problem. Two international conventions, The London Convention on the Prevention of Marine Pollution by Dumping of Wastes and Other Matter (IMO, 1972), and the MARPOL (1973/1978) Convention for the Prevention of Pollution from Ships (IMO, 1973/1978) have been designed to ensure that ship-sourced wastes are not dumped into the open ocean, but are provided with adequate ports reception facilities for the safe and efficient discharge and treatment of the wastes. In particular, Annex V (Pollution by Garbage from Ships) to the MARPOL convention, which came into force in 1989, specifically bans the dumping of persistent wastes, including plastics, at sea, and requests for countries to operate adequate ports reception facilities that should include garbage reception boats, ships and vehicles, garbage treatment facilities and adequate communication services, amongst others.

Parties and signatories to the MARPOL convention have been provided with adequate guidelines for operating sustainable and efficient national port reception facilities and guidance for ships to practice good housekeeping onboard (Hwang, 2020; IMO, 2013, 2014, 2016; Wallace & Coe, 1998). Unfortunately, whilst in most high-income countries, national ports authorities provide efficient port reception facilities (Argüello, 2020; NOWPAP, 2009; Øhlenschlæger et al., 2013), the same cannot be said of most seaports in Africa. African seaports have been characterised by their inability to provide adequate and sustainable infrastructure and suffer from generally poor management to address the issues of Annex V of the MARPOL Convention. Although the exact amounts of marine litter derived from port and harbour activities have not been quantified yet in Africa, the few available studies on the status of reception facilities at some national ports underscore the need for improvement. For example, several studies examining facilities at the Apapa Port and the TinCan Island Port of Lagos, Nigeria (Onwuegbuchunam et al., 2017a, 2017b; Osaloni, 2019; Peters & Marvis, 2019) have identified the need for improvements to meet MARPOL Convention requirements. In South Africa, a detailed audit of eight major ports, in Durban, Richards Bay, Cape Town, Saldanha, Ngqura, Port Elizabeth, East London and Mossel Bay, and eleven smaller ones (APWC, 2020) found that it was challenging to get a clear picture of the management of ship-generated waste received at commercial ports. Port-generated waste was well managed and regulated, but ship-generated waste had much lower levels of control. A study of ports in the Mediterranean region, including those of Algeria, Egypt, Morocco, Tunisia, and Libya (REMPEC, 2005), also highlighted the need for considerable improvement of port reception facilities in the African countries of that region. The story is not much different in the East African region (Lane et al., 2007).

Various port activities have also been linked with the unintentional release of microplastics into the environment. In 2017 an accidental spill occurred in Durban Harbour, South Africa, wherein two containers carrying PE pellets broke open after falling off a vessel. This resulted in the rapid and widespread distribution of pellets across the South African coastline (Schumann et al., 2019). Although clean-up campaigns were initiated about 82% of the pellets lost were never recovered (Schumann et al., 2019), probably due to seepage into the south Atlantic and the Indian Oceans via dominant ocean currents (Collins & Hermes, 2019) (Fig. 1.1a). Hull scrapings and marine coatings have also been identified as sources of marine microlitter as a result of harbour and port activities (Dibke et al., 2021). Interestingly, Preston-Whyte et al. (2021) suggested harbour/port dredge spoils that may get dumped in nearby coastal zones may be an important and understudied source of microplastics to the marine environment. This is especially important as many harbours in Africa are associated with high microplastic concentrations and a more diverse suite of microplastic particles (Chouchene et al., 2019; Naidoo et al., 2015; Nel et al., 2017; Shabaka et al., 2019). Harbour sediments are also highly contaminated with persistent organic pollutants (POPs), metals, and a whole range of other hazardous substances that have been shown to sorb to microplastics (Torres et al., 2021). Please refer to Chap. 1 for details on these other chemical pollutants.

2.2.2 Sea-Based Sources

Shipping Industry

As mentioned in the previous section, the shipping industry is a significant contributor to marine pollution including, dumping hazardous and general waste (Ryan et al., 2019). A long-term study of bottles washing ashore of Inaccessible Island in the South Atlantic Ocean found an increase in the debris of Asian origin; this suggested that dumping from ships played a significant role in marine pollution in that region (Ryan et al., 2019). Additionally, Ryan (2020a) investigated the origin of plastic bottles stranded on nine Kenyan beaches and concluded that most bottles in urban areas were from local sources. The presence of newly manufactured Polyethylene terephthalate (PET) bottles from China implied ship derived waste is still an important component. Further evidence of ship-based waste was observed in South Africa, wherein foreign bottles accounted for up to 74% on some beaches (Ryan et al., 2021).

Fishing and Aquaculture Industry

A meta-analysis of 68 publications estimated that annual losses of fishing nets, traps, and lines are around 6%, 9%, and 29%, respectively (Richardson et al., 2019). An analysis of the Great Pacific Garbage Patch estimated that fishing nets accounted for 46% of the mass of all plastics (Lebreton et al., 2018) whilst fishing related debris was commonly observed on the seabed across all areas of the West European coastal shelves (Maes et al., 2018).

In Africa, fishing gear has also been found to be a major contributor to marine litter from coastal to oceanic waters (Alshawafi et al., 2017; Loulad et al., 2017; Ryan, 2014; Scheren et al., 2002) and on the seabed (Ryan et al., 2020c; Woodall et al., 2015). In Morocco, plastic fishing gear accounted for 94% by number of all collected plastic items on the seafloor (Loulad et al., 2017). Whereas, in South Africa, fishery waste accounted for 22% by number and 73% by mass (Ryan et al., 2020c). In cases where fishing gear has been located close to shore, this is mainly related to small-scale fishing operations. Cardoso and Caldeira (2021) investigated the source of plastic pollution found on the Macaronesian Islands. They concluded that the high proportion of fishing gear littering Cabo Verde (Aguilera et al., 2018) originates predominantly from activities off Western Sahara, Mauritanian and Senegalese coasts, with the east coast of North America a secondary source (Cardoso & Caldeira, 2021).

Aquaculture has also been identified as a potential source of plastic into the marine environment through discarded or lost gear (e.g., polyvinyl chloride (PVC) tubes, nets, and cages) as a result of mismanagement and/or accidental losses during extreme weather events (Huntington, 2019). Aquaculture is also a potential source of microplastics leakage resulting from the fragmentation of plastic gear over time and contaminated fishmeal (Lusher et al., 2017; Thiele et al., 2021; Zhou et al., 2020a). The extent to which aquaculture contributes to marine pollution has not been a research focus in Africa, which may be attributed to the industry's small size (see details in Chap. 1). However, as African aquaculture is projected to increase (see Chap. 1, Fig. 1.2a, b), this could be an increasing source of marine litter.

Oil and Gas Industry

There is global concern about the contribution of the oil and gas industry to plastic pollution, especially at sea (Ahmed et al., 2021). Studies are limited however, a case study from the Norwegian continental shelf found higher microplastic concentrations in both sediment and tube-dwelling polychaete worms near offshore oil and gas installations compared to more remote reference sites (Knutsen et al., 2020). The European Union has recently commissioned studies on identifying all material inputs and activities of this industry that may contribute to the environmental burden of microplastics. It is already well known that microplastics are used in the following applications in the oil and gas sector: cement additives and loss circulation material for drilling, wax inhibitors in production, and crosslinking chemicals in pipelines (Amec Foster Wheeler Environment & Infrastructure UK Limited, 2017). Mega and macroplastic leakage from the oil and gas industry is unquantified. Current Oil-producing, coastal African countries (Algeria, Angola, Republic of Congo, Egypt, Equatorial Guinea, Gabon, Libya, Nigeria, and recently Ghana) often have offshore, and coast-based oil industry installations and are liable to release plastics of all sizes into the marine environment. Though oil production is currently concentrated on the north and west coasts of Africa, oil fields, which may be exploited in the future, do exist on the east coast. Gas-producing coastal countries occur all around Africa and are also liable to release plastics of all sizes into the marine environment. However, studies and data quantifying releases from this industry have been sparse despite this knowledge, in Africa and globally.

2.3   Abundance and Distribution of Marine Litter

The abundance, accumulation rates, and characteristics of marine litter can be investigated in different environmental compartments (shorelines, in coastal waters, and the open ocean from the surface to the seabed) using various methods. Watercourses (e.g., rivers, stormwater drains, and WWTP outlets) are also an area of interest as they provide a conduit for litter transportation into the marine environment. Fluxes between compartments are dynamic, and sinks can become sources to other areas and vice versa depending on various abiotic and biotic processes. As such, it is important to understand these fluxes between land and sea, from rivers, estuaries, and the nearshore surface water to the deep sea, to interpret data trends accurately. More importantly, Ryan et al. (2020b) suggest that when monitoring the effectiveness of mitigation measures, sites close to sources are better as they give a more rapid and accurate measure.

Models used to estimate current, and future risk scenarios predict flow and accumulation. These are often associated with high levels of uncertainty, especially due to an incomplete understanding of marine litter inputs and distribution processes and limited and incomparable datasets (Lebreton et al., 2017; Schmidt et al., 2017). A good demonstration of the uncertainty associated with estimating plastic flows into the marine environment is the case of South Africa. Based on estimates of total waste production and proportion of mismanaged waste, Jambeck et al. (2015) estimated that 90,000–250,000 tonnes of plastic flowed into the ocean from South Africa during 2010. A subsequent estimate by Verster and Bouwman (2020), using more robust data, put forth a more conservative range of 15,000–40,000 tonnes per year, highlighting that the amount of plastic flowing into the environment, though less, is still a point of concern. This last estimate was better aligned with observed amounts of plastic washing up on beaches (Weideman et al., 2020b). Global studies using a Lagrangian model have attempted to estimate marine litter hotspots, suggesting the Mediterranean Sea and the coastal zone around southern Africa as regions of concern (Eriksen et al., 2014; Lebreton et al., 2012; Van Sebille et al., 2015). Models can also assist with where litter has originated. For example, Van Der Mheen et al. (2020) investigated the distribution patterns of particles released into the Northern Indian Ocean (NIO). They suggested that depending on the particle beaching probability, the east coast of Africa and many SIDs can be severely affected by pollution released by countries in south-east Asia. This is supported by direct evidence of long-distance drift of high-density-PE bottles and lids, mainly from Indonesia, found on beaches in Kenya, South Africa, and various western Indian Ocean (WIO) island states (Duhec et al., 2015; Ryan, 2020a; Ryan et al., 2021).

2.3.1 Rivers

Freshwater environments are also contaminated with litter, with rivers considered a major pathway for land-based litter to enter the marine environment (Schmidt et al., 2017; Van Calcar & Van Emmerik, 2019). River basins can also retain high levels of litter (buried beneath sediment, trapped along rocky outcrops and vegetated areas), particularly during low-flow conditions. There have been a handful of studies in Africa looking at macrolitter associated with rivers. In South Africa, visual observations of litter flowing down three rivers into Algoa Bay estimated discharge rates of 22–1500 items day^{-1} (Moss et al., 2021). Weideman et al. (2020c) investigated the long-distance transport of litter within the Orange-Vaal River system and found limited downstream distribution, with macrolitter often linked to local sources. Ryan and Perold (2021) showed limited debris dispersion from a river into the ocean, with deposition concentrated on beaches within 1 km of the river mouth. They also observed the litter exchange between the sea and the river, with marine litter, found up to 1.2 km inland. Rivers also can be long-term sinks for litter (Ryan & Perold, 2021; Tramoy et al., 2020) however, this is dependent on climatic and hydrological conditions within the catchment. For more information on monitoring litter in rivers and lakes, please see the UNEP (2020) report on harmonised approaches.

Rivers as a major transportation pathway for litter has made it a key point of focus for intervention efforts. Several catchment litter management options exist (Armitage & Rooseboom, 2000), using river booms as a popular intervention method (Box 2.2).

Box 2.2: The Litterboom Project, South Africa

Interception booms made of sun-proof high-density PE have been placed in series across several inland rivers in Durban and Cape Town, collecting a minimum of 14,000 kg per site annually. The litter booms are designed to float on the water's surface, catching floating plastic and other debris as they move downstream, bound for the open ocean. The booms are placed at an angle to ensure waste flows towards the most accessible bank for easier and safer collection. Litter is recycled where possible or landfilled. Litter booms are easy to maintain for teams in the community and very effective when cleared daily. However, retention can be poor when flow rates are high, for example during rainfall events. Such projects have been useful also for raising community awareness on the impact of indiscriminate disposal and littering. In addition, the collected waste data can be used to inform city-wide efforts to stop ocean-bound plastics (Image 2.2).

Image 2.2 The Litterboom Project (Photo Credit). The initiative is credited to the partnership of The Litterboom Project (TLC) with Parley, the City of Cape Town and Pristine Earth Collective

Micro and nano-plastic and fibre abundances have also been assessed in some freshwater rivers across the continent (Alimi et al., 2021), with most studies in South Africa (Dahms et al., 2020; Dalu et al., 2021; Nel et al., 2018) and Nigeria (Adeogun et al., 2020; Akindele et al., 2019; Ebere et al., 2019; Oni et al., 2020). Microplastics were generally partitioned into the water, river bed/bank sediment, and biota. They were characterised to be mostly derived from PE, PP, PU, polystyrene (PS), and polyester materials (Alimi et al., 2021).

Microplastic abundance in inland freshwater systems across the continent is very varied. For the Bloukrans River system in the Eastern Cape of South Africa, Nel et al. (2018) found sediment microplastic concentrations were less in summer (6.3 ± 4.3 particles kg^{-1}) than in winter (160 ± 140 particles kg^{-1}). In Tunisia, Toumi et al. (2019) investigated the sediments of the Bizerte Lagoon and surrounding areas and found 2340–6920 particles kg^{-1} in streams, and 3000–18,000 particles kg^{-1} in the lagoon. For Lake Victoria, Egressa et al. (2020) found 0.02–2.19 particles m^{-3} in the water. For the same lake, in Kenya, Migwi et al. (2020) found 1.56–5.38 particles m^{-3} in the water. Concentrations are often difficult to compare directly due to different authors' variable sampling and analysis methodologies with no standard or harmonised approach available to date.

What drives microplastic distribution, immobilisation and remobilisation, and burial in freshwater systems is still understudied. Depending on various in-stream abiotic and biotic processes, these particles may become temporarily immobilised within riverbed sediments and other in-stream features or float freely within the water column (Drummond et al., 2020; Krause et al., 2021). Floating particles may get distributed further downstream, eventually discharging into the marine environment (Besseling et al., 2017; Drummond et al., 2020; Schmidt et al., 2017; Siegfried et al., 2017). Overall, there are data gaps regarding the extent African rivers contribute litter to marine ecosystems, whether this contribution varies seasonally and how future scenarios may change with the changing climate.

Other data gaps surround where the potentially vulnerable ecosystems are due to litter accumulation. Wetlands, for example, may be potential sinks for both land- and sea-derived litter (Ryan & Perold, 2021). An assessment of macrolitter in two mangrove forests in Mauritius observed mean densities of 0.46 ± 0.24 and 0.24 ± 0.22 items m^{-2} (Seeruttun et al., 2021). Additionally, microplastics have been detected in South African mangroves at densities ranging from 18.5 ± 34.4 per 500 g (St. Lucia) to 143.5 ± 93.0 per 500 g (Isipingo estuary) for sediment samples (Govender et al., 2020). Microplastics in water, sediment, and biota were also found associated with the coastal wetland of Sakumo II Lagoon in Ghana (Kanhai et al., 2017). Mangroves situated within 20 km of river mouths are more vulnerable to plastic pollution due to their potential to trap receiving litter (Harris et al., 2021). However, the extent these regions play as litter traps has not yet been established empirically.

2.3.2 Urban Drainage Systems

Few studies have been conducted on urban drainage systems in Africa, including stormwater drains and sewage outlets. Stormwater runoff resulting from rainfall events carries litter from many sources (Image 2.3), flushing debris into streams, rivers, and ultimately the sea. When stormwater occurs around coastal areas, beach litter may be directly washed into the sea. Most of Sub-Saharan Africa experiences stormwater events during the rainy seasons. In South Africa, Arnold and Ryan (1999) quantified urban stormwater runoff in Cape Town, observing macrolitter fluxes of 7–731 items ha^{-1} day^{-1}. Twenty years later, Weideman et al. (2020a), repeated the study finding little change with fluxes of 5–576 items ha^{-1} day^{-1}.

Stormwater also carries microlitter deposited from a variety of sources, including fragmented solid waste, city dust, tyre and road wear particles, paint chips, and other industrial and agricultural emissions (Boucher & Friot, 2017; Horton & Dixon, 2018;

Image 2.3 Stormwater drain discharge from Cape Town, South Africa (Photo Credit: T.Y. Chitaka) and stormwater debris deposited at the drainage entrance into the Sierra Leone River in Freetown (Photo Credit: S.K. Sankoh)

Liu et al., 2019; Pramanik et al., 2020). Stormwater runoff was considered a major pathway for microlitter to enter aquatic environments and has been shown as an important point source in Durban Harbour, South Africa (Preston-Whyte et al., 2021). Particles released from tyres, and brake pads constitute a major global source of microplastic contamination (Järlskog et al., 2020; Klöckner et al., 2020; Knight et al., 2020; Kole et al., 2017). Evangeliou et al. (2020) estimated that about 64,000 tonnes per year of tyre wear and brake wear particles are directly transported globally through rivers into the ocean, whilst about 140,000 tonnes per year are carried through long-range transport in the atmosphere and deposited into the sea. However, current extraction and spectroscopic techniques used to isolate and identify microplastics are often inadequate for tyre and road wear particle detection (Baensch-Baltruschat et al., 2020). Another source may come from the use of plastic waste in construction and infrastructure, such as roads, that may release plastic fragments over time (Appiah et al., 2017). With the growing economy of many African countries and the significant rise in the number of automobiles in use, some African cities are likely contributing significantly to the local contamination of the environment by microplastics from tyre and brake pad wear.

Runoff from agricultural land may be another pathway by which microlitter enters aquatic environments. Agricultural land may receive microlitter from the degradation of shade cloth, the application of contaminated sewage sludge or biosolids, use of slow-release plastic-encapsulated fertilisers, plastic mulch film, polymer coasted seeds, contaminated irrigation water, and from direct atmospheric deposition to farmland (Katsumi et al., 2021; Okoffo et al., 2021; Qi et al., 2020; Weithmann et al., 2018). Runoff can transport microplastics from farmlands into drainage systems and river courses. Wind can also mobilise soil-deposited microplastics into the atmosphere (Dris et al., 2016; Zhang et al., 2020), which can be especially important in arid zones such as the Sahara Desert where stormwater events are rare, and the wind frequently generates sandstorms that may be transported far beyond the immediate region.

2.3.3 Beaches

Beach litter surveys are the most common monitoring employed in the marine environment. Data gathered are often used to provide initial insight into the composition and quantity of marine litter and to infer the source. Most beach surveys in Africa have been conducted in South Africa, accounting for about 40% of all published studies (Table 2.1). However, the last 20 years have seen studies conducted in Algeria, Ghana, Guinea-Bissau, Kenya, Mauritania, Morocco, Nigeria, Seychelles, Tanzania, and Tunisia.

Beach litter surveys, using a transect or quadrats, are popular for two reasons; beaches are more accessible than other compartments (e.g., rocky shores, deep-sea, and open ocean) and require fewer resources (Barnardo & Ribbink, 2020, Annex 2.2). Furthermore, beach litter surveys also contribute to awareness-raising

Table 2.1 Review of observed litter densities across various beaches in Africa for different size fractions (macro, meso, and micro)

Country	Source	Survey date	Survey type	Material type	Size fraction	Observed density	Site #
Algeria	Tata et al. (2020)	2017/2018	Standing stock	Plastic	Micro and macro	649 ± 184 and 183 ± 27.32 kg^{-1} dry sediment	4
	Taïbi et al. (2021)	2017	Standing stock	Plastic	Micro and macro	7.6 ± 18.8 and 66 ± 107 items m^{-2} sediment	9
Ghana	Nunoo and Quayson (2003)	2002	Weekly accumulation	All	Macro	698 ± 62.99 and 876 ± 79.93 items 1000 m^{-2} week 7253 ± 618 and 5951 ± 783 g 1000 m^{-2} week	2
Bijagós archipelago, Guinea-Bissau	Lourenço et al. (2017)	2013/2015	Standing stock	Fibres	Micro	2.7 ± 3.27 fibres ml^{-1}	1
Kenya	Okuku et al. (2020a)	2019	Standing stock	All	Meso	$68–613.6$ items m^{-2}	23
	Okuku et al. (2020b)	2019	Daily accumulation	All	Macro	$3.8 \pm 3.1–24.9 \pm 19.1$ items m^{-1} day^{-1} $0.31 \pm 0.2–0.04 \pm 0.02$ g m^{-1} day^{-1}	6
	Okuku et al. (2021b)	2019–2020	Standing stock	All	Macro	$0.091–0.736$ items m^{-2}	1
Mauritania	Lourenço et al. (2017)	2013/2015	Standing stock	Fibres	Micro	4.3 ± 4.90 fibres ml^{-1}	1
Morocco	Velez et al. (2019)	Unknown	Standing stock	Plastic	Micro	Mean: 336 particles m^{-2}	
			Standing stock	All	Macro	$0.81 \pm 0.56–12.48 \pm 4.35$ items 5 m^{-1}	4
	Nachite et al. (2019)	2015–2017	Standing stock	All	Macro	$126 \pm 54.4–821 \pm 306$ items 100 m^{-1} $0.011 \pm 0.005–0.103 \pm 0.038$ g m^{-2}	14
	Haddout et al. (2021)	2020	Standing stock	Plastic	Micro	$10–300$ particles kg^{-1} sediment	18

(continued)

Table 2.1 (continued)

Country	Source	Survey date	Survey type	Material type	Size fraction	Observed density	Site #
Nigeria	Ilechukwu et al. (2019)	2018	Standing stock	Plastic	Micro	170 ± 21–121 ± 38 items 50 g^{-1} dry sediment	4
	Fred-Ahmadu et al. (2020)	2018–2019	Standing stock	Plastic	Micro	3.6 ± 3.5–173 ± 21.3 particles kg^{-1} sediment	5
Seychelles	Dunlop et al. (2020)	2003–2019	Daily accumulation rate	All	All	0.0010–0.0415 items m^{-1} day^{-1}	1
South Africa	Ryan and Moloney (1990)	1984	Standing stock	All	Micro	491 particles m^{-1}	52
					Macro	1.09 items m^{-1}	
		1989			Micro	678 particles m^{-1}	
					Macro	2.99 items m^{-1}	
	Madzena and Lasiak (1997)	1994	Standing stock	All	All	19.6–72.5 items m^{-1} 42.8–164.1 g m^{-1}	6
		1994–1995	Monthly accumulation			1.4–9.8 items m^{-1} month^{-1} 3.4–25.0 g m^{-1} month^{-1}	
	Ryan et al. (2014a)	1994	Daily accumulation	All	Macro	1.55 and 0.35 items m^{-1} day^{-1}	2
			Weekly accumulation			0.46 and 0.16 items m^{-1} day^{-1}	
		1995	Daily accumulation			2.87 and 1.30 items m^{-1} day^{-1}	
			Weekly accumulation			0.93 and 0.62 items m^{-1} day^{-1}	

(continued)

Table 2.1 (continued)

Country	Source	Survey date	Survey type	Material type	Size fraction	Observed density	Site #
	Nel and Froneman (2015)	2012	Daily accumulation	Plastic	Micro	14.58 and 2.04 items m^{-1} day^{-1}	21
			Weekly accumulation			5.47 and 0.63 items m^{-1} day^{-1}	
		2014	Standing stock	Plastic	Micro	689 ± 348–3308 ± 1449 particles m^{-2} sediment	
	Ryan et al. (2018)	2015	Standing stock	All	Meso	708 items m^{-1}	82
	Nel et al. (2017)	2016	Standing stock	Plastic	Micro	86.67 ± 48.68–755 ± 393 particles m^{-2}	13
	De Villiers (2018)	2016	Standing stock	Fibres	Micro	4–772 fibres dm^{-3} sediment	175
		2017				0–797 fibres dm^{-3} sediment	175
	Chitaka (2020)	2017	Daily accumulation	All	Macro	37.8–2962 items 100 m^{-1} day^{-1} 189–4430 g 100 m^{-1} day^{-1}	5
		2018–19			Macro	305–2082 items 100 m^{-1} day^{-1} 557–3799 g 100 m^{-1} day^{-1}	
	Ryan et al. (2020d)	2008	Standing stock	All	Macro	Surface: 11.8 items m^{-1}, 249 g m^{-1} Buried: 123.2 items m^{-1}, 149.0 g m^{-1}	1
		2010	Standing stock	All	Macro	Surface: 14.6 items m^{-1}, 227 g m^{-1} Buried: 92.8 items m^{-1}, 77.7 g m^{-1}	
				All	Meso	444 items m^{-1} 9.1 g m^{-1}	

(continued)

Table 2.1 (continued)

Country	Source	Survey date	Survey type	Material type	Size fraction	Observed density	Site #
		2017	Standing stock	Fibres	Micro	Surface: 60×10^3 fibres m^{-1}, 12.4 mg m^{-1} Buried: 128×10^3 fibres m^{-1}, 47.5 mg m^{-1}	
	Weideman et al. (2020b)	2019	Monthly accumulation rate	All	Macro	2.3 ± 0.8 items m^{-1} month^{-1} 8.5 ± 8.8 g m^{-1} month^{-1}	1
		2020	Daily accumulation rate			0.4 ± 0.3 items $100\ m^{-1}$ day 8.3 ± 20.4 g $100\ m^{-1}$ day^{-1}	2
Tanzania	Mayoma et al. (2020)	–	Standing stock	Plastic	Micro	15 ± 4–2972 ± 238 particles kg^{-1} dry weight	18
	Maione (2021)	2018	Standing stock	Plastic	Macro	46.8–89.7 kg	4
Tunisia	Abidli et al. (2018)	2017	Standing stock	Plastic	Micro	141 ± 25.98 and 461 ± 29.74 items kg^{-1} dry weight	5
	Missawi et al. (2020)	2018	Standing stock	Plastic	Micro	129–606 items kg^{-1} sediment	8

and positive behaviour change of those involved (Nelms et al., 2017). There are two general methods; standing stock surveys or accumulation rate surveys, with the latter currently only conducted for macrolitter. Standing stock surveys report the amount of litter at a specific period in time whilst the latter reports the accumulation rate of litter in a given area and can be used as a proxy for litter abundance in adjacent coastal waters subject to inputs from direct littering or exhumation (Cheshire et al., 2009; Ryan et al., 2009). When interpreting the results of beach surveys, it is important to consider the limitations of each method (see Annex 2.2 for further details). Macrolitter monitoring at beaches might be useful to determine the most prevalent items and subsequent actions, for comparable monitoring of status and trends of beach litter across countries and regions. It is also recommended to focus on the larger microplastic fraction (2–5 mm) as a potential legacy contaminant concentrations are likely to increase in the future (Chubarenko et al., 2020; Haseler et al., 2018).

Long-term longitudinal studies of standing stocks may provide indications of gross changes in the types and abundance of litter, as well as distribution patterns (Ryan et al., 2009). For example, Ryan et al. (2018) used a series of surveys conducted across South Africa in 1994, 2005, and 2015 to investigate mesoplastic distribution patterns, concluding that they mostly derive from local, land-based sources. In addition, there was no significant change in mesoplastic abundance over the years. In Kenya, Okuku et al. (2021b) employed standing stock surveys to investigate the influence of monsoons on the abundance and distribution of macrolitter in Mkomani Beach; the results indicated that monsoons influenced both litter abundance and composition.

Accumulation rates are highly site-specific, with variability across beaches and within beaches (Table 2.1). In 2019, accumulation rates of 3.8 ± 3.1–24.9 ± 19.1 items m^{-1} day^{-1} were observed in Kenya (Okuku et al., 2020b), whilst 0.0255 ± 0.0086 items m^{-1} day^{-1} were observed on Cousine Island, Seychelles (Dunlop et al., 2020) and 0.403 ± 0.061–0.853 ± 0.085 items m^{-1} day^{-1} in South Africa (Chitaka & von Blottnitz, 2019). Limited long-term studies investigating litter fluxes have been conducted. In South Africa, Ryan et al. (2014a) conducted daily and weekly accumulation rate surveys over two beaches in 1994, 1995, and 2012, during which a significant increase was observed in litter loads over time. On Cousine Island, Seychelles, Dunlop et al. (2020) conducted what is arguably the longest temporal study of litter fluxes in Africa, conducting 40 surveys from 2003 to 2019 along the same beach, significant increase in litter was observed over time.

To fully appreciate the extent of the marine litter problem, it is important to relate it to waste generation. A study in Cape Town, South Africa, estimated the proportion of products that leaked into the marine environment in 2017. It was found that items associated with food consumed on the go were more prone to leakage (Chitaka & von Blottnitz, 2021). The estimates were based on beach accumulation rates as a proxy for litter flows into the ocean, coupled with waste generation rates. Whilst uncertainty is associated with such estimates, it is important to note the differences in leakage rates for specific product items (Fig. 2.3).

2.3.4 Coastal and Oceanic Waters

Marine litter has been detected in coastal and oceanic waters off the African coast. However, the presence of marine litter in the ocean remains one of the most understudied compartments from an African perspective, as illustrated in Fig. 2.1a, b.

Seabed trawls conducted in the Mediterranean Sea, off the coast of Morocco, found total macrolitter densities ranging from 0 to 1768 ± 298 kg km^{-2} at depths up to 266 m (Loulad et al., 2017). Off the South African coast, only 17% of 235 trawls contained litter with an average density of 3.4 items km^{-2}. Most litter was located at depths greater than 200 m (Ryan et al., 2020c). From 2012 to 2015, Loulad et al. (2019) conducted sea trawl surveys in the Mediterranean Sea and observed mean densities of 26 ± 68–80 ± 133 kg km^{-2}, most of which was located closer to the coast. Visual surveys conducted in the South Atlantic Ocean in 2013 observed a decrease in macrolitter density as distance increased from the coast of Cape Town (Ryan, 2014). Furthermore, the survey offered the first evidence of a South Atlantic 'garbage patch'. Subsequent surveys provided further evidence to support the dispersion and accumulation of litter into this gyre (Ryan et al., 2014b).

Product leakage rates in Cape Town in 2017

LOW PREVALENCE

Percentage

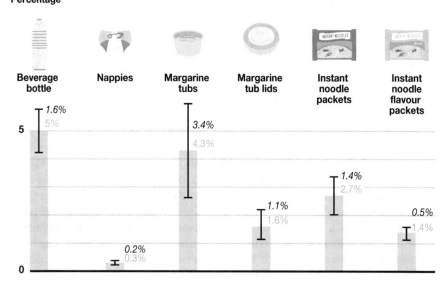

Source: Chitaka and von Blottnitz, 2021.

GRID-Arendal/Studio Atlantis, 2021

Fig. 2.3 Looking at the big picture: product-specific leakage rates

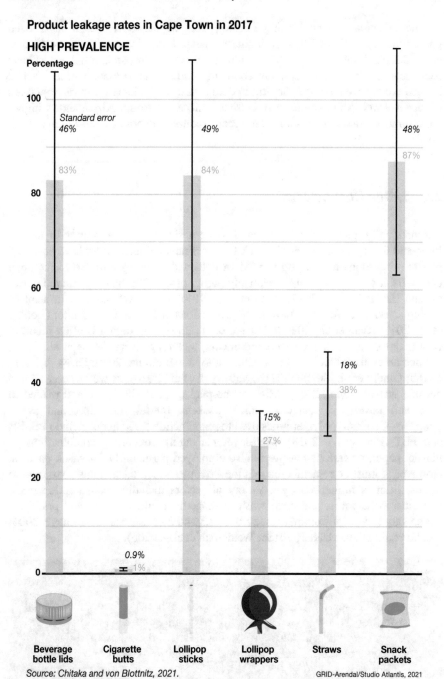

Product leakage rates in Cape Town in 2017

HIGH PREVALENCE

Percentage

Source: Chitaka and von Blottnitz, 2021.

GRID-Arendal/Studio Atlantis, 2021

Fig. 2.3 (continued)

Microplastics collected using neuston nets in the open ocean have been predominately made of PE, with higher concentrations closer to the coast (Suaria et al., 2020b; Vilakati et al., 2020). Fibrous microlitter (rayon and polyester) have been detected in seabed sediment south of Madagascar (Woodall et al., 2015). Seabed cores have also been helpful in demonstrating the increase in microplastics in recent years. An example from Durban Harbour in South Africa shows higher concentrations associated with more recent sediment deposits (Matsuguma et al., 2017).

2.4   Litter Characteristics

Internationally, plastic has been found to be a significant contributor to marine litter, and this is the same case in Africa. Assessments of macrolitter have observed plastic proportions ranging up to 99% of collected items by number. In the open ocean, fishing gear makes up a relatively higher proportion than is observed closer to land (Loulad et al., 2017; Ryan et al., 2014b). Moreover, areas with a lot of fishing activity are found to have large proportions of fishing-related litter (Loulad et al., 2017; Ryan et al., 2020d; Scheren et al., 2002). Packaging is often found to be a major contributor across compartments, including rivers (Moss et al., 2021; Weideman et al., 2020c), beaches (Chitaka & von Blottnitz, 2019; Fazey & Ryan, 2016b; Okuku et al., 2020b, 2021b; Van Dyck et al., 2016), as well as coastal and oceanic waters (Ryan, 2014). Most of the packaging is single-use and related to food and beverages, including snack packets, bottles, lids/caps, and sweet wrappers. Common polymer types used to manufacture these items include PE, PP, and PET (PlasticsSA, 2018). Multilayer packaging containing combinations of plastic, paper, or various plastics is also employed particularly for snack packets. However, it must be remembered that the extent to which these items contribute to marine litter is influenced by a variety of factors including consumption rates, consumer behaviours, and solid waste management infrastructure and practices; which vary across the continent (see Boxes 2.3 and 2.4) (Marais & Armitage, 2004; Okuku et al., 2020b; UNEP, 2018a; Weideman et al., 2020b).

Box 2.3: The Scourge of Water Sachets in West Africa

For several decades, the West Africa sub-region has been bedevilled by a special form of plastic waste—sachets used for packaging water, which now serves most people with a safe source of drinking water. It began as an initiative by a local entrepreneur in Nigeria in the 1990s and has grown into a lucrative business throughout West Africa. Its rapid growth stems from the failure of governments to provide clean and safe potable water and sanitation (GIZ, 2019; Stoler, 2017; Thomas et al., 2020; WWAP, 2015).

The water is packed into low-density-PE sachets holding 300–500 ml. National regulations on the production, use, and management of waste derived from this product are largely disregarded by manufacturers and consumers and are not enforced (Vapnek & Williams, 2017). A 'use and throw away' culture generally still prevails in the region, resulting in massive littering of street corners, open drains, and streams and rivers. Water sachets are amongst the top contributors to beach litter in Ghana (Nunoo & Quayson, 2003; Tsagbey et al., 2009) and Nigeria (Ebere et al., 2019).

Polymer identification is a very important step in meso and microplastic research as it is used to infer the most likely source/origin, as well as suggest associated risk/hazard. As the source is easier to infer when litter is larger, polymer identification techniques are not regularly employed during macrolitter studies. The lack of spectroscopy equipment hinders the African continents ability to identify polymers (Akindele & Alimba, 2021; Alimi et al., 2021; Nel et al., 2021). In South Africa, the application of the rapid screening technique using a fluorescent dye (Nile Red) (Maes et al., 2017) has proved to be a cost-effective solution for the large scale monitoring of microplastics in marine sediment, water, and fish (Bakir et al., 2020; Preston-Whyte et al., 2021). Shabaka et al. (2019) used differential scanning calorimetry (DSC) to detect a wide range of polymers in Eastern Harbour, Egypt; detecting PP, polyethylene vinyl acetate (PEVA), acrylonitrile butadiene, PS, and polytetrafluoroethylene. There also appears to be capacity using Attenuated Total Reflection-Fourier Transform Infrared (ATR-FTIR) Spectroscopy for the identification of particles > 300 μm (Nel et al., 2021). However, it would be pertinent to build capacity for polymer identification < 300 μm, given that this size range is often linked to increased uptake (Chap. 3).

2.4.1 Factors Influencing Litter Characteristics, Abundance, and Distribution

Several factors have been identified to determine the characteristics, abundance, and distribution of litter in the marine environment. These include, but are not limited to:

- Catchment area characteristics and drainage systems
- Development status and income levels of residents
- Climatic condition (wind, rainfall amount, and flood events)
- Coastal hydrodynamics and ocean currents
- Physical and chemical characteristics of the litter materials.

Catchment Area Characteristics and Drainage System

Once litter is in the marine environment, several processes can influence characteristics, abundance, distribution, and fate. Catchment area characteristics (land-use cover, population density) have influenced litter. In South Africa, Arnold and Ryan (1999) and Weideman et al. (2020a) conducted studies quantifying macrolitter discharges from the same three catchment areas (residential, industrial, and mixed commercial/residential) during wetter months in 1996 and 2018–2019. In both cases, macrolitter was most abundant in the industrial area, with the least in the residential area. In general, remote locations are associated with lower macrolitter abundance than those in densely populated areas (Nachite et al., 2019; Okuku et al., 2020b; Ryan, 2020a; Seeruttun et al., 2021). Nevertheless, some remote beaches away from industrial/urban centres have been found to have relatively high litter loads, which suggest long-range transportation does occur (Aguilera et al., 2018; Dunlop et al., 2020; Ryan et al., 2019). In addition, a study by Ryan et al. (2021) found that standing stocks at remote beaches had lower bottle loads than urban beaches but higher loads than semi-urban beaches; this was attributed to the lower inputs of semi-urban beaches vs urban, coupled with greater cleaning efforts at semi-urban beaches than remote beaches.

Development Status and Income Levels of Residents

Some studies have suggested an inverse relationship between income level and macrolitter abundance. A study of debris in stormwater drains in South Africa found higher macrolitter loads in low-income areas, which was attributed to the poor waste removal services available (Marais & Armitage, 2004). A similar relationship was suggested by accumulation surveys of five beaches conducted in Cape Town, wherein a beach in a low-income area was associated with relatively high macrolitter loads (Chitaka & von Blottnitz, 2019).

Climatic Condition (Wind, Rainfall Amount, and Flood Events)

Litter distribution is influenced by climatic conditions such as wind and rain. Rainfall and flood events can increase litter fluxes from watercourses and waterways as accumulated litter is flushed out of the system (Chitaka & von Blottnitz, 2021; Nunoo & Quayson, 2003; Okuku et al., 2020b; Ryan & Perold, 2021). Wind strength and direction have also influenced litter distribution and deposition (Okuku et al., 2021b; Ryan & Perold, 2021).

Coastal Hydrodynamics and Ocean Currents

Ocean currents play a vital role in transporting and distributing litter within the marine environment (Collins & Hermes, 2019; Van Sebille et al., 2015). Trawl surveys have observed variations in litter density according to water depth (Loulad et al., 2017, 2019; Ryan et al., 2020c). There are no clear correlation indicating if this variation is direct (lower depth, less litter) or inverse (lower depth, more litter) variation. Litter distribution within the ocean is also influenced by geomorphology and hydrodynamics (Loulad et al., 2019), additionally, ocean currents influence the deposition of litter on coastlines (Collins & Hermes, 2019; Ryan, 2020b). A study conducted by Chitaka and von Blottnitz (2019) suggested preferential deposition of litter in Table Bay (South Africa) which was attributed to water movements. Further studies on litter deposition along South African coastlines have also suggested that water movements significantly influence the distribution and stranding of litter items (Fazey & Ryan, 2016a, 2016b; Ryan & Perold, 2021). Microplastic distribution and their fate are influenced by ocean currents (Collins & Hermes, 2019; Schumann et al., 2019), biofouling and inclusion within sinking marine snow (Kooi et al., 2017), sequestration along deep-sea canyons (Pohl et al., 2020) and fluxes to the atmosphere via sea breeze (Allen et al., 2020).

Physical Characteristics of the Litter Materials

The physical characteristics of an item also influence distribution of litter. Fazey and Ryan (2016a, 2016b) found that size and buoyancy influence debris dispersal, with smaller and less buoyant items observed to disperse over shorter distances. Biofouling was also found to play a role in distribution by decreasing the buoyancy of items (Fazey & Ryan, 2016a). Furthermore, Weideman et al. (2020b) found that rigid plastics were less likely to be deposited and trapped along rocky shorelines, whilst flexible packaging was prone to entrapment in weeds and rocks.

Similar to larger litter items, microplastic distribution is linked to particle size and shape, polymer type, density, surface characteristics, and degradation rates, to name a few. These factors may affect which types of particles the marine environment receives through river inputs. Weideman et al. (2020c) found that fibres were present across the Orange-Vaal river basin but concentrated in the lower reaches. At the same time large plastics and fragments were more closely linked to urban settlements. Chouchene et al. (2019) recorded features indicative of weathering (i.e., pits, fractures, grooves, cracks, and scratches) associated with PE and PP microplastics from Sidi Mansour Harbour sediment samples in Tunisia. There is a need to understand how factors, such as weathering, influence the transport and fate of microplastics. Changes to plastics (bites on bottles, biofouling) can also be used as an indicator of the length of time plastic litter has been at sea and potentially travelled (Ryan, 2020a; Ryan et al., 2021).

Fibres appear more homogenously distributed within the environment (Barrows et al., 2018; Ryan et al., 2020d; Weideman et al., 2019, 2020c). This may reflect their multiple entry points, for example, point sources through WWTP effluent, and diffuse sources via atmospheric deposition and through the spread of sewage sludge or biosolids on agricultural land. Alternatively, fibres may be more widely distributed, especially as their large surface area to volume ratio may lead to reduced settling rates compared to other microplastic shapes (Hoellein et al., 2019; Khatmullina & Isachenko, 2017). More research is needed to corroborate this assumption, using controlled lab-based experiments, such as artificial flumes which can test hydrodynamic scenarios. Visual bias may also lead to conspicuous fibres being detected more frequently than other microlitter types however, few studies have tested this empirically. Nevertheless, it is important for researchers to understand what type of microplastics are being transported down rivers to the marine environment and in what 'condition'; as they have likely undergone a series of immobilisation and remobilisation events changing their physical and chemical characteristics that in turn will change how they behave in the estuarine and marine environment as they are no longer 'pristine'. Understanding these processes within the global, let alone the African context is still in its infancy.

Some microplastics and microfibres emitted from different sources are present and suspended in the atmosphere (Dris et al., 2016) and liable to long-range transport to remote places, including parts of Africa (Evangeliou et al., 2020; González-Pleiter et al., 2021; Wright et al., 2020; Zhang et al., 2020). However, no data is published for Africa regarding atmospheric contamination by microplastics and microfibres. Microplastics get deposited onto soil surfaces by gravitational settling and rainout/washout processes during wet precipitation events (Brahney et al., 2020). This is an substantial gap to fill, especially considering atmospheric deposition may be a major diffuse source for aquatic and terrestrial environments (Wright et al., 2020; Zhang et al., 2020).

Box 2.4: Litter and the COVID-19 Pandemic

The COVID-19 pandemic highlighted the usefulness of plastic in our society in the form of Personal Protective Equipment. Unfortunately, the increased consumption of single-use plastics and their improper disposal raised concerns about the impacts on the environment. In Kenya, Okuku et al. (2021a) found that COVID-19 related litter, including masks, gloves, soap wrappers, wet wipes as well as liquid hand wash and sanitiser bottles, were observed along 11 of 14 streets 10-days after the first confirmed COVID-19 case in the country, contributing up to 17% of waste items. In comparison, in South Africa, relatively low amounts of COVID-19 related litter were observed during daily accumulation surveys of urban streets, contributing less than 1% (Ryan et al., 2020a). On Kenyan beaches, COVID-19 litter densities of up to 5.6×10^{-2} items m^{-2} were observed (Okuku et al., 2021a). Interestingly, higher densities were observed at remote beaches, attributed to less compliance with Government instructions to close beaches. This complements findings by Ryan et al. (2020a), who found approximately three times as much litter during less restricted periods of national lockdown compared to periods where movement was strictly monitored. Additionally, the potential of masks as a source of microplastics has been suggested by some researchers (Fadare & Okoffo, 2020; Shruti et al., 2020).

2.5 Key Messages and Future Directions

This chapter demonstrates that current published studies are isolated to a few selected countries. Thus, there is a need for more coordinated research efforts, using harmonised approaches, across Africa. Specifically, it is important to develop baseline datasets which when combined with long-term monitoring studies at the same locales, will enable countries to measure change and mitigation effectiveness. This can only be realised through investment in capacity across the continent, especially equipment and expertise at the smaller size fractions of plastic pollution (micro and nano), which may have legacy impacts.

Knowledge of the pathways and sources for litter release in the environment can facilitate concentrated mitigation efforts and aid in the accurate interpretation of monitoring datasets in the future. There is a need for more field studies quantifying litter inputs, across all size ranges, from various sources. For example, WWTPs are an understudied source of plastics into the environment in Africa, whilst landfill and incineration sites may be important legacy sources in the future if not contained effectively. Rivers are a pathway for the transportation and transformation of plastics however, understanding the role small and larger systems play in retaining plastics is also important for risk-based assessments, clean-up efforts, and interpretation of downstream trends. This will require inter-African collaboration, especially as many rivers are transboundary. Many large and important rivers in Africa, i.e., the Nile, Congo, Niger, Zambezi, and Senegal, have not been extensively studied.

More studies looking at distribution and burial drivers, and underlying fragmentation processes are required. This will enable a deeper understanding of the results of monitoring studies such as beach surveys. For example, the need to assess the role seasonality plays in litter distribution and burial. Hurley et al. (2018) showed that seasonal changes in catchment hydrology could redistribute microlitter hotspots. This needs to be done across the continent as wet and dry seasons will be regionally relevant and can vary significantly within countries and across the continent. This is especially important as climate change is expected to alter the duration and intensity of various climatic events that could change how litter is immobilised and remobilised in the environment. Other aspects such as the occurrence/degree of biofouling and fragmentation of litter and the hetero-aggregation of micro and nano-plastics may vary with the different current and future climatic conditions found across Africa.

Models looking at how litter gets distributed from urban-industrial centres around Africa or the numerous rivers discharging into the marine environment are important in understanding marine pollution. However, this can only be achieved through various actions such as;

- Hosting workshops whereby researchers working on in situ data collection and those who need data for model validation are gathered to discuss what is required for models versus what is available/achievable.
- The development of an open access database on plastic pollution abundance/loads specific to the African continent.
- Capacity building for more modelling expertise in Africa.

Whilst research is essential to developing an understanding of plastic pollution; this is not to imply that countries should postpone developing strategies to mitigate litter inputs. It is also vital to understand the drivers of littering and inappropriate waste management (with a view to more effectively changing adverse behaviours). We know there is a problem, and efforts should be made to mitigate it by developing product targeted interventions taking into account the leakage propensities of different items (Fig. 2.3a, b). Thus, combining accumulation rate studies with waste generation rates to get a fuller picture of leakage rates into the environment should be encouraged.

Acknowledgements We would like to acknowledge the valuable insights provided by Peter Ryan, Anham Salyani, Salieu Sankoh, Tony Ribbink, and Elvis Okoffo in peer-reviewing this chapter. We would also like to acknowledge Nieves López and Federico Labanti (Studio Atlantis) for creating the illustrations and Cameron Service for sharing information relating to the the Litterboom Project. Thanks to Dr T. Chitaka, Dr S. Sankoh and the Litterboom Project for allowing the use of their photographs.

Annex 2.1: Marine Litter Quantification Studies Published Across Africa in Peer-Reviewed Journals as of December 2021

Country	Total number of studies	Citations
Ghana	7	Scheren et al. (2002), Nunoo and Quayson (2003), Tsagbey et al. (2009), Hosoda et al. (2014), Van Dyck et al. (2016), Chico-Ortiz et al. (2020), Gbogbo et al. (2020)
Cote D'Ivoire	1	Scheren et al. (2002)

(continued)

(continued)

Country	Total number of studies	Citations
Benin	1	Scheren et al. (2002)
Cameroon	1	Scheren et al. (2002)
Nigeria	2	Scheren et al. (2002), Ebere et al., (2019)
Kenya	6	Kosore et al. (2018), Ryan (2020a), Okuku et al. (2020a, 2020b, 2021a, 2021b)
Cousine Island, Seychelles	1	Dunlop et al. (2020)
Alphonse Island, Seychelles	1	Duhec et al. (2015)
Mauritius	1	Seeruttun et al. (2021)
Morocco	7	Alshawafi et al. (2017), Loulad et al. (2017), Maziane et al. (2018), Nachite et al. (2019), Velez et al. (2019), Mghili et al. (2020), Haddout et al. (2021)
South Africa	37	Ryan (1988, 2015, 2020b), Ryan and Moloney (1990), Madzena and Lasiak (1997), Ryan et al. (2014a, 2018, 2020a, 2020b, 2020c, 2020d, 2021), Naidoo et al. (2015), Nel and Froneman (2015), Fazey and Ryan (2016), Matsuguma et al. (2017), Nel et al. (2017, 2018, 2021), De Villiers (2018, 2019), Chitaka and von Blottnitz (2019), Collins and Hermes (2019), Naidoo and Glassom (2019), Schumann et al. (2019), Weideman et al. (2019, 2020a, 2020b, 2020c), Govender et al. (2020), Iroegbu et al. (2020), Vetrimurugan et al. (2020), Verster and Bouwman (2020), Vilakati et al. (2020), Moss et al. (2021), Preston-Whyte et al. (2021), Ryan and Perold (2021)
Algeria	2	Mankou-Haddadi et al. (2021), Taïbi et al. (2021)
Tunisia	5	Chouchene et al. (2019, 2020), Missawi et al. (2020), Wakkaf et al. (2020), Zayen et al. (2020)
Tanzania	2	Mayoma et al. (2020), Maione (2021)
Egypt	1	Shabaka et al. (2019)
Mauritania	1	Lourenço et al. (2017)
Guinea-Bissau	1	Lourenço et al. (2017)
Senegal	1	Tavares et al. (2020)
Atlantic Ocean	4	Ryan (2014), Kanhai et al. (2017), Ryan et al. (2019, 2020b)

(continued)

(continued)

Country	Total number of studies	Citations
Southern Ocean	2	Ryan et al. (2014b), Suaria et al. (2020)
Mediterranean	2	Cózar et al. (2015), Cincinelli et al. (2019)
Indian Ocean	2	Woodall et al. (2014), Connan et al. (2021)

References for Annex 2.1

Alshawafi, A., Analla, M., Alwashali, E., & Aksissou, M. (2017). Assessment of marine debris on the coastal wetland of Martil in the North-East of Morocco. *Marine Pollution Bulletin, 117*, 302–310. https://doi.org/10.1016/j.marpolbul.2017.01.079

Chico-Ortiz, N., Mahu, E., Crane, R., Gordon, C., & Marchant, R. (2020). Microplastics in Ghanaian coastal lagoon sediments: Their occurrence and spatial distribution. *Regional Studies in Marine Science, 40*, 101509. https://doi.org/10.1016/j.rsma.2020.101509

Chitaka, T. Y., & von Blottnitz, H. (2019). Accumulation and characteristics of plastic debris along five beaches in Cape Town. *Marine Pollution Bulletin, 138*. https://doi.org/10.1016/j.marpolbul.2018.11.065

Chouchene, K., da Costa, J. P., Wali, A., Girão, A. V., Hentati, O., Duarte, A. C., et al. (2019). Microplastic pollution in the sediments of Sidi Mansour Harbor in Southeast Tunisia. *Marine Pollution Bulletin, 146*, 92–99. https://doi.org/10.1016/j.marpolbul.2019.06.004

Chouchene, K., Rocha-Santos, T., & Ksibi, M. (2020). Types, occurrence, and distribution of microplastics and metals contamination in sediments from south west of Kerkennah archipelago, Tunisia. *Environmental Science and Pollution Research*. https://doi.org/10.1007/s11356-020-09938-z

Cincinelli, A., Martellini, T., Guerranti, C., Scopetani, C., Chelazzi, D., & Giarrizzo, T. (2019). A potpourri of microplastics in the sea surface and water column of the Mediterranean Sea. *TrAC—Trends in Analytical Chemistry, 110*, 321–326. https://doi.org/10.1016/j.trac.2018.10.026

Collins, C., & Hermes, J. C. (2019). Modelling the accumulation and transport of floating marine micro-plastics around South Africa. *Marine Pollution Bulletin, 139*, 46–58. https://doi.org/10.1016/j.marpolbul.2018.12.028

Connan, M., Perold, V., Dilley, B. J., Barbraud, C., Cherel, Y., & Ryan, G. (2021). The Indian Ocean 'garbage patch': Empirical evidence from floating macro-litter. *Marine Pollution Bulletin, 169*. https://doi.org/10.1016/j.marpolbul.2021.112559

Cózar, A., Sanz-Martín, M., Martí, E., González-Gordillo, J. I., Ubeda, B., Gálvez, J. Á., et al. (2015). Plastic accumulation in the Mediterranean Sea. *PLOS One, 10*, 1–12. https://doi.org/10.1371/journal.pone.0121762

De Villiers, S. (2018). Quantification of microfibre levels in South Africa's beach sediments, and evaluation of spatial and temporal variability from 2016 to 2017. *Marine Pollution Bulletin, 135*, 481–489. https://doi.org/10.1016/j.marpolbul.2018.07.058

De Villiers, S. (2019). Microfibre pollution hotspots in river sediments adjacent to South Africa's coastline. *Water SA, 45*, 97–102. https://doi.org/10.4314/wsa.v45i1.11

Duhec, A. V., Jeanne, R. F., Maximenko, N., & Hafner, J. (2015). Composition and potential origin of marine debris stranded in the Western Indian Ocean on remote Alphonse Island, Seychelles. *Marine Pollution Bulletin, 96*, 76–86. https://doi.org/10.1016/j.marpolbul.2015.05.042

Dunlop, S. W., Dunlop, B. J., & Brown, M. (2020). Plastic pollution in paradise: Daily accumulation rates of marine litter on Cousine Island, Seychelles. *Marine Pollution Bulletin, 151*, 110803. https://doi.org/10.1016/j.marpolbul.2019.110803

Ebere, E. C., Wirnkor, V. A., Ngozi, V. E., & Chukwuemeka, I. S. (2019). Macrodebris and microplastics pollution in Nigeria: First report on abundance, distribution and composition. *Environmental Analysis, Health and Toxicology, 34*, e2019012. https://doi.org/10.5620/eaht.e2019012

Fazey, F. M. C., & Ryan, P. G. (2016). Debris size and buoyancy influence the dispersal distance of stranded litter. *Marine Pollution Bulletin, 110*, 371–377. https://doi.org/10.1016/j.marpolbul.2016.06.039

Gbogbo, F., Takyi, J. B., Billah, M. K., & Ewool, J. (2020). Analysis of microplastics in wetland samples from coastal Ghana using the Rose Bengal stain. *Environmental Monitoring and Assessment, 192*. https://doi.org/10.1007/s10661-020-8175-8

Govender, J., Naidoo, T., Rajkaran, A., Cebekhulu, S., Bhugeloo, A., & Sershen (2020). Towards characterising microplastic abundance, typology and retention in Mangrove-dominated estuaries. *Water, 12*, 2802. https://doi.org/10.3390/w12102802

Haddout, S., Gimiliani, G. T., Priya, K. L., Hoguane, A. M., Casila, J. C. C., & Ljubenkov, I. (2021). Microplastics in surface waters and sediments in the Sebou Estuary and Atlantic Coast, Morocco. *Analytical Letters,* 1–13. https://doi.org/10.1080/00032719.2021.1924767

Hosoda, J., Ofosu-Anim, J., Sabi, E. B., Akita, L. G., Onwona-Agyeman, S., Yamashita, R., et al. (2014). Monitoring of organic micropollutants in Ghana by combination of pellet watch with sediment analysis: E-waste as a source of PCBs. *Marine Pollution Bulletin, 86*, 575–581. https://doi.org/10.1016/j.marpolbul.2014.06.008

Iroegbu, A. O. C., Sadiku, R. E., Ray, S. S., & Hamam, Y. (2020). Plastics in municipal drinking water and wastewater treatment plant effluents: Challenges and opportunities for South Africa—A review. *Environmental Science and Pollution Research, 27*, 12953–12966. https://doi.org/10.1007/s11356-020-08194-5

Kanhai, L. D. K., Officer, R., Lyashevsha, O., Thompson, R. C., & O'Connor, I. (2017). Microplastic abundance, distribution and composition along a latitudinal gradient in the Atlantic Ocean. *Marine Pollution Bulletin, 115*(1), 307–314. https://doi.org/10.1016/j.marpolbul.2016.12.025

Kosore, C., Ojwang, L., Maghanga, J., Kamau, J., Kimeli, A., Omukoto, J., et al. (2018). Occurrence and ingestion of microplastics by zooplankton in Kenya's marine environment: First documented evidence. *African Journal of Marine Science, 40*, 225–234. https://doi.org/10.2989/1814232X.2018.1492969.

Loulad, S., Houssa, R., Rhinane, H., Boumaaz, A., & Benazzouz, A. (2017). Spatial distribution of marine debris on the seafloor of Moroccan waters. *Marine Pollution Bulletin, 124*, 303–313. https://doi.org/10.1016/j.marpolbul.2017.07.022

Lourenço, P. M., Serra-Gonçalves, C., Ferreira, J. L., Catry, T., & Granadeiro, J. P. (2017). Plastic and other microfibers in sediments, macroinvertebrates and shorebirds from three intertidal

wetlands of southern Europe and West Africa. *Environmental Pollution, 231*, 123–133. https://
doi.org/10.1016/j.envpol.2017.07.103

Madzena, A., & Lasiak, T. (1997). Spatial and temporal variations in beach litter on the Transkei
coast of South Africa. *Marine Pollution Bulletin, 34*, 900–907. https://doi.org/10.1016/S0025-
326X(97)00052-0

Maione, C. (2021). Quantifying plastics waste accumulations on coastal tourism sites in Zanzibar,
Tanzania. *Marine Pollution Bulletin, 168*, 112418. https://doi.org/10.1016/j.marpolbul.2021.
112418

Mankou-Haddadi, N., Bachir-bey, M., Galgani, F., Mokrane, K., & Sidi, H. (2021). Benthic marine
litter in the coastal zone of Bejaia (Algeria) as indicators of anthropogenic pollution. *Marine
Pollution Bulletin, 170*, 112634. https://doi.org/10.1016/j.marpolbul.2021.112634

Matsuguma, Y., Takada, H., Kumata, H., Kanke, H., Sakurai, S., Suzuki, T., et al. (2017).
Microplastics in sediment cores from Asia and Africa as indicators of temporal trends in
plastic pollution. *Archives of Environmental Contamination and Toxicology, 73*, 230–239.
https://doi.org/10.1007/s00244-017-0414-9

Mayoma, B. S., Sørensen, C., Shashoua, Y., & Khan, F. R. (2020). Microplastics in beach
sediments and cockles (*Anadara antiquata*) along the Tanzanian coastline. *Bulletin of
Environmental Contamination and Toxicology, 105*, 513–521. https://doi.org/10.1007/s00128-
020-02991-x

Maziane, F., Nachite, D., & Anfuso, G. (2018). Artificial polymer materials debris characteristics
along the Moroccan Mediterranean coast. *Marine Pollution Bulletin, 128*, 1–7. https://doi.org/
10.1016/j.marpolbul.2017.12.067

Mghili, B., Analla, M., Aksissou, M., & Aissa, C. (2020). Marine debris in Moroccan Mediterranean
beaches: An assessment of their abundance, composition and sources. *Marine Pollution Bulletin,
160*, 111692. https://doi.org/10.1016/j.marpolbul.2020.111692

Missawi, O., Bousserrhine, N., Belbekhouche, S., Zitouni, N., Alphonse, V., Boughattas, I., et al.
(2020). Abundance and distribution of small microplastics ($\leq 3\,\mu m$) in sediments and seaworms
from the Southern Mediterranean coasts and characterisation of their potential harmful effects.
Environmental Pollution, 263, 114634. https://doi.org/10.1016/j.envpol.2020.114634

Moss, K., Allen, D., González-Fernández, D., & Allen, S. (2021). Filling in the knowledge gap:
Observing macroplastic litter in South Africa's rivers. *Marine Pollution Bulletin, 162*. https://
doi.org/10.1016/j.marpolbul.2020.111876

Nachite, D., Maziane, F., Anfuso, G., & Williams, A. T. (2019). Spatial and temporal variations of
litter at the Mediterranean beaches of Morocco mainly due to beach users. *Ocean & Coastal
Management, 179*, 104846. https://doi.org/10.1016/j.ocecoaman.2019.104846

Naidoo, T., & Glassom, D. (2019). Sea-surface microplastic concentrations along the coastal shelf
of KwaZulu–Natal, South Africa. *Marine Pollution Bulletin, 149*, 110514. https://doi.org/10.
1016/j.marpolbul.2019.110514

Naidoo, T., Glassom, D., & Smit, A. J. (2015). Plastic pollution in five urban estuaries of KwaZulu-
Natal, South Africa. *Marine Pollution Bulletin, 101*, 473–480. https://doi.org/10.1016/j.marpol
bul.2015.09.044

Nel, H. A., Chetwynd, A. J., Kelly, C. A., Stark, C., Valsami-Jones, E., Krause, S., et al. (2021). An
untargeted thermogravimetric analysis-Fourier transform infrared-gas chromatography–mass
spectrometry approach for plastic polymer identification. *Environmental Science & Technology,
55*(13), 8721–8729. https://doi.org/10.1021/acs.jpca.1c07162

Nel, H. A., Dalu, T., & Wasserman, R. J. (2018). Sinks and sources: Assessing microplastic
abundance in river sediment and deposit feeders in an Austral temperate urban river system.
Science of the Total Environment, 612, 950–956. https://doi.org/10.1016/j.scitotenv.2017.
08.298

Nel, H. A., & Froneman, P. W. (2015). A quantitative analysis of microplastic pollution along the south-eastern coastline of South Africa. *Marine Pollution Bulletin, 101*, 274–279. https://doi. org/10.1016/j.marpolbul.2015.09.043

Nel, H. A., Hean, J. W., Noundou, X. S., & Froneman, P. W. (2017). Do microplastic loads reflect the population demographics along the southern African coastline? *Marine Pollution Bulletin, 115*, 115–119. https://doi.org/10.1016/j.marpolbul.2016.11.056

Nunoo, F. K. E., & Quayson, E. (2003). Towards management of litter accumulation—Case study of two beaches in Accra, Ghana. *Journal of the Ghana Science Association, 5*, 145–155.

Okuku, E. O., Kiteresi, L. I., Owato, G., Mwalugha, C., Omire, J., Mbuche, M., et al. (2020a). Baseline meso-litter pollution in selected coastal beaches of Kenya: Where do we concentrate our intervention efforts? *Marine Pollution Bulletin, 158*, 111420. https://doi.org/10.1016/j.mar polbul.2020.111420

Okuku, E. O., Kiteresi, L. I., Owato, G., Mwalugha, C., Omire, J., Otieno, K., et al. (2020b). Marine macro-litter composition and distribution along the Kenyan Coast: The first-ever documented study. *Marine Pollution Bulletin, 159*, 111497. https://doi.org/10.1016/j.marpolbul.2020.111497

Okuku, E., Kiteresi, L., Owato, G., Otieno, K., Mwalugha, C., Mbuche, M., et al. (2021a). The impacts of COVID-19 pandemic on marine litter pollution along the Kenyan Coast: A synthesis after 100 days following the first reported case in Kenya. *Marine Pollution Bulletin, 162*, 111840. https://doi.org/10.1016/j.marpolbul.2020.111840

Okuku, E. O., Kiteresi, L., Owato, G., Otieno, K., Omire, J., Kombo, M. M., et al. (2021b). Temporal trends of marine litter in a tropical recreational beach: A case study of Mkomani beach, Kenya. *Marine Pollution Bulletin, 167*, 112273. https://doi.org/10.1016/j.marpolbul.2021.112273

Preston-Whyte, F., Silburn, B., Meakins, B., Bakir, A., Pillay, K., Worship, M., et al. (2021). Meso- and microplastics monitoring in harbour environments: A case study for the Port of Durban, South Africa. *Marine Pollution Bulletin, 163*, 111948. https://doi.org/10.1016/j.marpolbul.2020. 111948

Ryan, P. G. (1988). The characteristics and distribution of plastic particles at the sea-surface off the southwestern Cape Province, South Africa. *Marine Environmental Research, 25*(4), 249–273.

Ryan, P. G. (2014). Litter survey detects the South Atlantic "garbage patch". *Marine Pollution Bulletin, 79*, 220–224. https://doi.org/10.1016/j.marpolbul.2013.12.010

Ryan, P. G. (2015). Does size and buoyancy affect the long-distance transport of floating debris? *Environmental Research Letters, 10*. https://doi.org/10.1088/1748-9326/10/8/084019

Ryan, P. G. (2020a). Land or sea? What bottles tell us about the origins of beach litter in Kenya. *Waste Management, 116*, 49–57. https://doi.org/10.1016/j.wasman.2020.07.044

Ryan, P. G. (2020b). The transport and fate of marine plastics in South Africa and adjacent oceans. *South African Journal of Science, 116*, 1–9. https://doi.org/10.17159/sajs.2020/7677

Ryan, P. G., Dilley, B. J., Ronconi, R. A., & Connan, M. (2019). Rapid increase in Asian bottles in the South Atlantic Ocean indicates major debris inputs from ships. *Proceedings of the National Academy of Sciences*, 1–6. https://doi.org/10.1073/pnas.1909816116

Ryan, P. G., Lamprecht, A., Swanepoel, D., & Moloney, C. L. (2014a). The effect of fine-scale sampling frequency on estimates of beach litter accumulation. *Marine Pollution Bulletin, 88*, 249–254. https://doi.org/10.1016/j.marpolbul.2014.08.036

Ryan, P. G., & Moloney, C. L. (1990). Plastic and other artefacts on South African beaches: Temporal trends in abundance and composition. *South African Journal of Science, 86*, 450–452.

Ryan, P. G., Musker, S., & Rink, A. (2014b). Low densities of drifting litter in the African sector of the Southern Ocean. *Marine Pollution Bulletin, 89*, 16–19. https://doi.org/10.1016/j.marpol bul.2014.10.043

Ryan, P. G., & Perold, V. (2021). Limited dispersal of riverine litter onto nearby beaches during rainfall events. *Estuarine, Coastal and Shelf Science, 251*, 107186. https://doi.org/10.1016/j. ecss.2021.107186

Ryan, P. G., Perold, V., Osborne, A., & Moloney, C. L. (2018). Consistent patterns of debris on South African beaches indicate that industrial pellets and other mesoplastic items mostly derive from local sources. *Environmental Pollution, 238,* 1008–1016. https://doi.org/10.1016/j.envpol.2018.02.017

Ryan, P. G., Pichegru, L., Perold, V., & Moloney, C. L. (2020a). Monitoring marine plastics—Will we know if we are making a difference? *South African Journal of Science, 116,* 1–9. https://doi.org/10.17159/sajs.2020/7678

Ryan, P. G., Suaria, G., Perold, V., Pierucci, A., Bornman, T. G., & Aliani, S. (2020b). Sampling microfibres at the sea surface: The effects of mesh size, sample volume and water depth. *Environmental Pollution, 258,* 113413. https://doi.org/10.1016/j.envpol.2019.113413

Ryan, P. G., Weideman, E. A., Perold, V., Durholtz, D., & Fairweather, T. P. (2020c). A trawl survey of seafloor macrolitter on the South African continental shelf. *Marine Pollution Bulletin, 150,* 110741. https://doi.org/10.1016/j.marpolbul.2019.110741

Ryan, P. G., Weideman, E. A., Perold, V., & Moloney, C. L. (2020d). Toward balancing the budget: Surface macro-plastics dominate the mass of particulate pollution stranded on beaches. *Frontiers in Marine Science, 7,* 1–14. https://doi.org/10.3389/fmars.2020.575395

Ryan, P. G., Weideman, E. A., Perold, V., Hofmeyr, G., & Connan, M. (2021). Message in a bottle: Assessing the sources and origins of beach litter to tackle marine pollution. *Environmental Pollution, 288,* 117729. https://doi.org/10.1016/j.envpol.2021.117729

Scheren, P. A., Ibe, A. C., Janssen, F. J., & Lemmens, A. M. (2002). Environmental pollution in the Gulf of Guinea—A regional approach. *Marine Pollution Bulletin, 44,* 633–641. https://doi.org/10.1016/S0025-326X(01)00305-8

Schumann, E. H., Fiona MacKay, C., & Strydom, N. A. (2019). Nurdle drifters around South Africa as indicators of ocean structures and dispersion. *South African Journal of Science, 115,* 1–9. https://doi.org/10.17159/sajs.2019/5372

Seeruttun, L. D., Raghbor, P., & Appadoo, C. (2021). First assessment of anthropogenic marine debris in mangrove forests of Mauritius, a small oceanic island. *Marine Pollution Bulletin, 164,* 112019. https://doi.org/10.1016/j.marpolbul.2021.112019

Shabaka, S. H., Ghobashy, M., & Marey, R. S. (2019). Identification of marine microplastics in Eastern Harbor, Mediterranean Coast of Egypt, using differential scanning calorimetry. *Marine Pollution Bulletin, 142,* 494–503. https://doi.org/10.1016/j.marpolbul.2019.03.062

Suaria, G., et al. (2020). Floating macro- and microplastics around the Southern Ocean: Results from the Antarctic circumnavigation expedition. *Environment International, 136,* 105494.

Taïbi, N. E., Bentaallah, M. E. A., Alomar, C., Compa, M., & Deudero, S. (2021). Micro- and macro-plastics in beach sediment of the Algerian western coast: First data on distribution, characterization, and source. *Marine Pollution Bulletin, 165,* 112168. https://doi.org/10.1016/j.marpolbul.2021.112168

Tavares, D. C., Moura, J. F., Ceesay, A., & Merico, A. (2020). Density and composition of surface and buried plastic debris in beaches of Senegal. *Science of the Total Environment, 737,* 139633. https://doi.org/10.1016/j.scitotenv.2020.139633

Tsagbey, S. A., Mensah, A. M., & Nunoo, F. K. E. (2009). Influence of tourist pressure on beach litter and microbial quality—Case study of two beach resorts in Ghana. *West African Journal of Applied Ecology, 15.* https://doi.org/10.4314/wajae.v15i1.49423

Van Dyck, I. P., Nunoo, F. K. E., & Lawson, E. T. (2016). An empirical assessment of marine debris, seawater quality and littering in Ghana. *Journal of Geoscience and Environment Protection, 04,* 21–36. https://doi.org/10.4236/gep.2016.45003

Velez, N., Zardi, G. I., Lo Savio, R., McQuaid, C. D., Valbusa, U., Sabour, B., et al. (2019). A baseline assessment of beach macrolitter and microplastics along northeastern Atlantic shores. *Marine Pollution Bulletin, 149,* 110649. https://doi.org/10.1016/j.marpolbul.2019.110649

Verster, C., & Bouwman, H. (2020). Land-based sources and pathways of marine plastics in a South African context. *South African Journal of Science, 116,* 1–9. https://doi.org/10.17159/sajs.2020/7700

Vetrimurugan, E., Jonathan, M. P., Sarkar, S. K., Rodríguez-González, F., Roy, P. D., Velumani, S., et al. (2020). Occurrence, distribution and provenance of micro plastics: A large scale quantitative analysis of beach sediments from southeastern coast of South Africa. *Science of the Total Environment, 746*, 141103. https://doi.org/10.1016/j.scitotenv.2020.141103

Vilakati, B., Sivasankar, V., Mamba, B. B., Omine, K., & Msagati, T. A. M. (2020). Characterization of plastic micro particles in the Atlantic Ocean seashore of Cape Town, South Africa and mass spectrometry analysis of pyrolyzate products. *Environmental Pollution, 265*, 114859. https://doi.org/10.1016/j.envpol.2020.114859

Wakkaf, T., El Zrelli, R., Kedzierski, M., Balti, R., Shaiek, M., Mansour, L., et al. (2020). Characterization of microplastics in the surface waters of an urban lagoon (Bizerte lagoon, Southern Mediterranean Sea): Composition, density, distribution, and influence of environmental factors. *Marine Pollution Bulletin, 160*, 111625. https://doi.org/10.1016/j.marpolbul.2020.111625

Weideman, E. A., Perold, V., Arnold, G., & Ryan, P. G. (2020a). Quantifying changes in litter loads in urban stormwater run-off from Cape Town, South Africa, over the last two decades. *Science of the Total Environment, 724*, 138310. https://doi.org/10.1016/j.scitotenv.2020.138310

Weideman, E. A., Perold, V., Omardien, A., Smyth, L. K., & Ryan, P. G. (2020b). Quantifying temporal trends in anthropogenic litter in a rocky intertidal habitat. *Marine Pollution Bulletin, 160*, 111543. https://doi.org/10.1016/j.marpolbul.2020.111543

Weideman, E. A., Perold, V., & Ryan, P. G. (2020c). Limited long-distance transport of plastic pollution by the Orange-Vaal River system, South Africa. *Science of the Total Environment, 727*, 138653. https://doi.org/10.1016/j.scitotenv.2020.138653

Weideman, E. A., Perold, V., & Ryan, P. G. (2019). Little evidence that dams in the Orange–Vaal River system trap floating microplastics or microfibres. *Marine Pollution Bulletin, 149*, 110664. https://doi.org/10.1016/j.marpolbul.2019.110664

Woodall, L. C., Sanchez-Vidal, A., Canals, M., Paterson, G. L. J., Coppock, R., Sleight, V., et al. The deep sea is a major sink for microplastic debris. *Royal Society Open Science, 1*(4), 140317. https://doi.org/10.1098/rsos.140317

Zayen, A., Sayadi, S., Chevalier, C., Boukthir, M., Ben Ismail, S., & Tedetti, M. (2020). Microplastics in surface waters of the Gulf of Gabes, southern Mediterranean Sea: Distribution, composition and influence of hydrodynamics. *Estuarine, Coastal and Shelf Science, 242*, 106832. https://doi.org/10.1016/j.ecss.2020.106832

Annex 2.2: Marine Litter Monitoring

Marine litter monitoring can be conducted for a number of reasons including changes in abundance and compositions of litter from different sources or in different compartments as well as assessing the effectiveness of mitigation efforts (Ryan et al., 2020). Beach surveys are a common method for monitoring marine litter due to the accessibility of the beaches compared to the open ocean or sea bed. Furthermore, relatively less equipment is required; personal protective equipment is required for participants, receptacles for collecting the litter and sieves for collecting small size fractions. They often focus on macrolitter, due to the difficulty associated with sampling smaller size fractions. Thus, the accessibility of this method makes it an attractive option for initial investigations into marine litter.

In general, either standing stock assessments or accumulation rate surveys are used. The former reports the amount of litter at a specific period in time whilst the latter reports the accumulation rate of litter in a given area. Both methods provide information on the abundance and characteristics of litter. Furthermore, accumulation rate surveys can be used to better understand litter fluxes between compartments (Cheshire et al., 2009; Ryan et al., 2009), whilst simultaneously giving a better reflection of overall standing stock associated with that location. For more details on monitoring refer to Barnardo and Ribbink (2020) and GESAMP (2019).

Standing stock surveys are popular as they are relatively less time intensive as they only require once-off sampling. However, as they provide a snapshot in time the information they provide with regards to marine litter is limited. More specifically, reported litter loads should be approached with caution as their representativeness and thus interpretation is constrained by the limited information regarding litter fluxes, distribution and deposition prior to the collection of litter. For example, an increase in standing stocks over fifty years can be attributed to a number of factors including an increase, decrease or even no change in litter washing ashore, turnover rates of different material types as well as beach cleaning efforts (Ryan et al., 2020). As such, the value of standing stock surveys lies in the litter composition observed rather than amounts.

Accumulation rate surveys are associated with greater investment in time and effort. They require an initial clean-up of the survey area followed by regular sampling of the newly arrived litter. Thus, they are better suited to macrolitter as it is difficult to ensure that smaller size fractions are completely collected during the initial clean-up (Ryan et al., 2020). Studies can be conducted at different intervals including, daily, weekly or monthly. However, observed fluxes are influenced by the chosen sampling frequency. A comparison of daily vs weekly sampling campaigns conducted by Ryan et al. (2014) found that daily surveys yielded 2.1–3.4 times more items than weekly, with observed masses 1.3–2.3 times greater. Furthermore, the study observed that low density items were associated with greater differences with polystyrene foam showing 4–5 times greater values during daily sampling. This demonstrated that different polymer types are associated with varying turnover rates, most likely linked to wind or perhaps to their buoyancy in the water column. In addition, observed accumulation rates can be influenced by water movements and climatic conditions including rain, wind strength and direction (Ryan et al., 2009). Other challenges include exhumation of buried litter either by tides, the weather or beach goers and cleaning efforts on the site (Ryan et al., 2020).

References for Annex 2.2

Barnardo, T., & Ribbink, A. J. (2020). *African marine litter monitoring manual.* African Marine Waste Network, Sustainable Seas Trust, Port Elizabeth. https://www.wiomsa.org/wp-content/uploads/2020/07/African-Marine-Litter-Monitoring-Manual_Final.pdf

Cheshire, A. C., Adler, E., Barbière, J., Cohen, Y., Evans, S., Jarayabhand, S., et al. (2009). *UNEP/IOC guidelines on survey and monitoring of marine litter* [UNEP Regional Seas Reports and Studies, No. 186; IOC Technical Series]. https://wedocs.unep.org/xmlui/bitstream/handle/20.500.11822/13604/rsrs186.pdf?sequence=1

GESAMP. (2019). *Guidelines or the monitoring and assessment of plastic litter and microplastics in the ocean.* IMO/FAO/UNESCO-IOC/UNIDO/WMO/IAEA/UN/UNEP/UNDP/ISA Joint Group of Experts on the Scientific Aspects of Marine Environmental Protection. http://www.gesamp.org/publications/guidelines-for-the-monitoring-and-assessment-of-plastic-litter-in-the-ocean

Ryan, P. G., Lamprecht, A., Swanepoel, D., & Moloney, C. L. (2014). The effect of fine-scale sampling frequency on estimates of beach litter accumulation. *Marine Pollution Bulletin, 88*, 249–254. https://doi.org/10.1016/j.marpolbul.2014.08.036

Ryan, P. G., Moore, C. J., Van Franeker, J. A., & Moloney, C. L. (2009). Monitoring the abundance of plastic debris in the marine environment. *Philosophical Transactions of the Royal Society B-Biological Sciences, 364*, 1999–2012. https://doi.org/10.1098/rstb.2008.0207

Ryan, P. G., Maclean, K., & Weideman, E. A. (2020a). The impact of the COVID-19 lockdown on urban street litter in South Africa. *Environmental Processes, 7*, 1303–1312. https://doi.org/10.1007/s40710-020-00472-1

Ryan, P. G., Pichegru, L., Perold, V., & Moloney, C. L. (2020b). Monitoring marine plastics—Will we know if we are making a difference? *South African Journal of Science, 116*, 58–66. https://sajs.co.za/article/view/7678/9944

Ryan, P. G., Weideman, E. A., Perold, V., Durholtz, D., & Fairweather, T. P. (2020c). A trawl survey of seafloor macrolitter on the South African continental shelf. *Marine Pollution Bulletin, 150*, 6. https://doi.org/10.1016/j.marpolbul.2019.110741

Ryan, P. G., Weideman, E. A., Perold, V., & Moloney, C. L. (2020d). Toward balancing the budget: Surface macro-plastics dominate the mass of particulate pollution stranded on beaches. *Frontiers in Marine Science, 7*, 14. https://doi.org/10.3389/fmars.2020.575395

References

Abidli, S., Antunes, J. C., Ferreira, J. L., Lahbib, Y., Sobral, P., & Trigui El Menif, N. (2018). Microplastics in sediments from the littoral zone of the north Tunisian coast (Mediterranean Sea). *Estuarine, Coastal and Shelf Science, 205*, 1–9. https://doi.org/10.1016/j.ecss.2018.03.006

Adeogun, A. O., Ibor, O. R., Khan, E. A., Chukwuka, A. V., Omogbemi, E. D., & Arukwe, A. (2020). Detection and occurrence of microplastics in the stomach of commercial fish species from a municipal water supply lake in southwestern Nigeria. *Environmental Science and Pollution Research, 27*, 31035–31045. https://doi.org/10.1007/s11356-020-09031-5

AfDB/UNEP/GRID-Arendal. (2020). *Sanitation and wastewater atlas of Africa.* AfDB/UNEP/Grid-Arendal. https://www.grida.no/publications/471

Aguilera, M., Medina-Suárez, M., Pinós, J., Liria-Loza, A., & Benejam, L. (2018). Marine debris as a barrier: Assessing the impacts on sea turtle hatchlings on their way to the ocean. *Marine Pollution Bulletin, 137*, 481–487. https://doi.org/10.1016/j.marpolbul.2018.10.054

Ahmed, F., Moodley, V., & Sookrajh, R. (2008). The environmental impacts of beach sport tourism events: A case study of the Mr Price Pro surfing event, Durban, South Africa. *Africa Insight, 38*, 73–85. https://www.cabdirect.org/cabdirect/abstract/20093084848

Ahmed, M. B., Rahman, M. S., Alom, J., Hasan, M. D. S., Johir, M. A. H., & Mondal, M. I. H. (2021). Microplastic particles in the aquatic environment: A systematic review. *Science of the Total Environment, 775*, 145793. https://doi.org/10.1016/j.scitotenv.2021.145793

Akindele, E. O., & Alimba, C. G. (2021). Plastic pollution threat in Africa: Current status and implications for aquatic ecosystem health. *Environmental Science and Pollution Research, 28*, 7636–7651. https://doi.org/10.1007/s11356-020-11736-6

Akindele, E. O., Ehlers, S. M., & Koop, J. H. E. (2019). First empirical study of freshwater microplastics in West Africa using gastropods from Nigeria as bioindicators. *Limnologica, 78*, 9. https://agris.fao.org/agris-search/search.do?recordID=US201900446650

Alimi, O. S., Fadare, O. O., & Okoffo, E. D. (2021). Microplastics in African ecosystems: Current knowledge, abundance, associated contaminants, techniques, and research needs. *Science of the Total Environment, 755*, 142422. https://doi.org/10.1016/j.scitotenv.2020.142422

Allen, S., Allen, D., Moss, K., Le Roux, G., Phoenix, V. R., & Sonke, J. E. (2020). Examination of the ocean as a source for atmospheric microplastics. *PLOS One, 15*, e0232746. https://doi.org/10.1371/journal.pone.0232746

Alshawafi, A., Analla, M., Alwashali, E., & Aksissou, M. (2017). Assessment of marine debris on the coastal wetland of Martil in the North-East of Morocco. *Marine Pollution Bulletin, 117*, 302–310. https://doi.org/10.1016/j.marpolbul.2017.01.079

Amec Foster Wheeler Environment & Infrastructure Uk Limited. (2017). *Intentionally added microplastics in products*. https://ec.europa.eu/environment/chemicals/reach/pdf/39168%20Intentionally%20added%20microplastics%20-%20Final%20report%2020171020.pdf

American Chemistry Council. (2021). https://www.opcleansweep.org/about/

Andriamahefazafy, M., & Failler, P. (2021). Towards a circular economy for African Islands: An analysis of existing baselines and strategies. *Circular Economy and Sustainability*. https://doi.org/10.1007/s43615-021-00059-4

Appiah, J. K., Berko-Boateng, V. N., & Tagbor, T. A. (2017). Use of waste plastic materials for road construction in Ghana. *Case Studies in Construction Materials, 6*, 1–7. https://doi.org/10.1016/j.cscm.2016.11.001

APWC. (2020). *Port reception waste facilities audit*. The Commonwealth Litter Programme. APWC (Asia Pacific Waste Consultants). https://www.cefas.co.uk/clip/resources/reports/africa-clip-reports/port-reception-waste-facilities-audit-south-africa-apwc/

Argüello, G. (2020). Environmentally sound management of ship wastes: Challenges and opportunities for European ports. *Journal of Shipping and Trade, 5*, 12. https://doi.org/10.1186/s41072-020-00068-w

Armitage, N., & Rooseboom, A. (2000). The removal of urban litter from stormwater conduits and streams: Paper 1—The quantities involved and catchment litter management options. *Water SA, 26*. https://www.researchgate.net/publication/267260648_The_removal_of_urban_litter_from_stormwater_conduits_and_streams_Paper_1_-The_quantities_involved_and_catchment_litter_management_options

Arnold, G., & Ryan, P. G. (1999). Marine litter originating from Cape Town's residential, commercial, and industrial areas: The stormwater connection.

Baensch-Baltruschat, B., Kocher, B., Stock, F., & Reifferscheid, G. (2020). Tyre and road wear particles (TRWP)—A review of generation, properties, emissions, human health risk, ecotoxicity, and fate in the environment. *Science of the Total Environment, 733*, 137823. https://doi.org/10.1016/j.scitotenv.2020.137823

Bakir, A., van der Lingen, C. D., Preston-Whyte, F., Bali, A., Geja, Y., Barry, J., Mdazuka, Y., Mooi, G., Doran, D., Tooley, F., Harmer, R., & Maes, T. (2020). Microplastics in commercially important small pelagic fish species from South Africa. *Frontiers in Marine Science, 7*, 910. https://doi.org/10.3389/fmars.2020.574663

Barnardo, T., & Ribbink, A. J. (2020). *African marine litter monitoring manual*. African Marine Waste Network, Sustainable Seas Trust. https://repository.oceanbestpractices.org/handle/11329/1420

Barrows, A. P. W., Christiansen, K. S., Bode, E. T., & Hoellein, T. J. (2018). A watershed-scale, citizen science approach to quantifying microplastic concentration in a mixed land-use river. *Water Research, 147*, 382–392. https://doi.org/10.1016/j.watres.2018.10.013

Besseling, E., Quik, J. T. K., Sun, M., & Koelmans, A. A. (2017). Fate of nano- and microplastic in freshwater systems: A modeling study. *Environmental Pollution, 220*, 540–548. https://doi.org/10.1016/j.envpol.2016.10.001

Bond, T., Ferrandiz-Mas, V., Felipe-Sotelo, M., & Van Sebille, E. (2018). The occurrence and degradation of aquatic plastic litter based on polymer physicochemical properties: A review. *Critical Reviews in Environmental Science and Technology, 48*, 685–722. https://doi.org/10.1080/10643389.2018.1483155

Boucher, J., & Friot, D. (2017). *Primary microplastics in the oceans: A global evaluation of sources.* IUCN. https://portals.iucn.org/library/sites/library/files/documents/2017-002-En.pdf

Brahney, J., Hallerud, M., Heim, E., Hahnenberger, M., & Sukumaran, S. (2020). Plastic rain in protected areas of the United States. *Science, 368*, 1257–1260. https://doi.org/10.1126/science.aaz5819

Browne, M. A., Ros, M., & Johnston, E. L. (2020). Pore-size and polymer affect the ability of filters for washing-machines to reduce domestic emissions of fibres to sewage. *PLOS One, 15*, e0234248. https://doi.org/10.1371/journal.pone.0234248

Bundhoo, Z. M. A. (2018). Solid waste management in least developed countries: Current status and challenges faced. *Journal of Material Cycles and Waste Management, 20*, 1867–1877. https://doi.org/10.1007/s10163-018-0728-3

Cardoso, C., & Caldeira, R. M. A. (2021). Modeling the exposure of the Macaronesia Islands (NE Atlantic) to marine plastic pollution. *Frontiers in Marine Science, 8*. https://doi.org/10.3389/fmars.2021.653502

Castaneda, R. A., Avlijas, S., Simard, M. A., & Ricciardi, A. (2014). Microplastic pollution in St. Lawrence River sediments. *Canadian Journal of Fisheries and Aquatic Sciences, 71*, 1767–1771. https://doi.org/10.1139/cjfas-2014-0281

Chan, C. K. M., Park, C., Chan, K. M., Mak, D. C. W., Fang, J. K. H., & Mitrano, D. M. (2021). Microplastic fibre releases from industrial wastewater effluent: A textile wet-processing mill in China. *Environmental Chemistry*. https://doi.org/10.1071/EN20143

Cheshire, A. C., Adler, E., Barbière, J., Cohen, Y., Evans, S., Jarayabhand, S., et al. (2009). *UNEP/IOC guidelines on survey and monitoring of marine litter* [UNEP Regional Seas Reports and Studies, No. 186; IOC Technical Series]. https://wedocs.unep.org/xmlui/bitstream/handle/20.500.11822/13604/rsrs186.pdf?sequence=1

Cheung, P. K., & Fok, L. (2017). Characterisation of plastic microbeads in facial scrubs and their estimated emissions in Mainland China. *Water Research, 122*, 53–61. https://doi.org/10.1016/j.watres.2017.05.053

Chitaka, T. Y. (2020). *Inclusion of leakage into life cycle management of products involving plastic as a material choice.* Faculty of Engineering and the Built Environment, Department of Chemical Engineering. http://hdl.handle.net/11427/32574

Chitaka, T. Y., & Von Blottnitz, H. (2019). Accumulation and characteristics of plastic debris along five beaches in Cape Town. *Marine Pollution Bulletin, 138*, 451–457. https://doi.org/10.1016/j.marpolbul.2018.11.065

Chitaka, T. Y., & Von Blottnitz, H. (2021). Development of a method for estimating product-specific leakage propensity and its inclusion into the life cycle management of plastic products. *The International Journal of Life Cycle Assessment*. https://doi.org/10.1007/s11367-021-01905-1

Chouchene, K., Da Costa, J. P., Wali, A., Girão, A. V., Hentati, O., Duarte, A. C., et al. (2019). Microplastic pollution in the sediments of Sidi Mansour Harbor in Southeast Tunisia. *Marine Pollution Bulletin, 146*, 92–99. https://doi.org/10.1016/j.marpolbul.2019.06.004

Chubarenko, I., Esiukova, E., Khatmullina, L., Lobchuk, O., Grave, A., Kileso, A., Haseler, M. (2020). From macro to micro, from patchy to uniform: Analyzing plastic contamination along and across a sandy tide-less coast. *Marine Pollution Bulletin, 156*. https://doi.org/10.1016/j.marpolbul.2020.111198

Collins, C., & Hermes, J. C. (2019). Modelling the accumulation and transport of floating marine micro-plastics around South Africa. *Marine Pollution Bulletin, 139,* 46–58. https://doi.org/10.1016/j.marpolbul.2018.12.028

Conley, K., Clum, A., Deepe, J., Lane, H., & Beckingham, B. (2019). Wastewater treatment plants as a source of microplastics to an urban estuary: Removal efficiencies and loading per capita over one year. *Water Research X, 3,* 100030. https://doi.org/10.1016/j.wroa.2019.100030

Cooper, D. A., & Corcoran, P. L. (2010). Effects of mechanical and chemical processes on the degradation of plastic beach debris on the island of Kauai, Hawaii. *Marine Pollution Bulletin, 60,* 650–654. https://doi.org/10.1016/j.marpolbul.2009.12.026

Dahms, H. T. J., Van Rensburg, G. J., & Greenfield, R. (2020). The microplastic profile of an urban African stream. *Science of the Total Environment, 731,* 138893. https://doi.org/10.1016/j.scitotenv.2020.138893

Dalu, T., Banda, T., Mutshekwa, T., Munyai, L. F., & Cuthbert, R. N. (2021). Effects of urbanisation and a wastewater treatment plant on microplastic densities along a subtropical river system. *Environmental Science and Pollution Research.* https://doi.org/10.1007/s11356-021-13185-1

Dauvergne, P. (2018). The power of environmental norms: Marine plastic pollution and the politics of microbeads. *Environmental Politics, 27,* 579–597. https://doi.org/10.1080/09644016.2018.1449090

De Falco, F., Cocca, M., Avella, M., & Thompson, R. C. (2020). Microfiber release to water, via laundering, and to air, via everyday use: A comparison between polyester clothing with differing textile parameters. *Environmental Science & Technology, 54,* 3288–3296. https://doi.org/10.1021/acs.est.9b06892

De Falco, F., Di Pace, E., Cocca, M., & Avella, M. (2019). The contribution of washing processes of synthetic clothes to microplastic pollution. *Scientific Reports, 9,* 6633. https://doi.org/10.1038/s41598-019-43023-x

De Villiers, S. (2018). Quantification of microfibre levels in South Africa's beach sediments, and evaluation of spatial and temporal variability from 2016 to 2017. *Marine Pollution Bulletin, 135,* 481–489. https://doi.org/10.1016/j.marpolbul.2018.07.058

De Villiers, S. (2019). Microfibre pollution hotspots in river sediments adjacent to South Africa's coastline. *Water SA, 45,* 97–102. https://doi.org/10.4314/wsa.v45i1.11

Dibke, C., Fischer, M., & Scholz-Böttcher, B. M. (2021). Microplastic mass concentrations and distribution in German bight waters by pyrolysis-gas chromatography–mass spectrometry/thermochemolysis reveal potential impact of marine coatings: Do ships leave skid marks? *Environmental Science & Technology, 55,* 2285–2295. https://doi.org/10.1021/acs.est.0c04522

Dris, R., Gasperi, J., Saad, M., Mirande, C., & Tassin, B. (2016). Synthetic fibers in atmospheric fallout: A source of microplastics in the environment? *Marine Pollution Bulletin, 104,* 290–293. https://doi.org/10.1016/j.marpolbul.2016.01.006

Drummond, J. D., Nel, H. A., Packman, A. I., & Krause, S. (2020). Significance of hyporheic exchange for predicting microplastic fate in rivers. *Environmental Science & Technology Letters, 7,* 727–732. https://doi.org/10.1021/acs.estlett.0c00595

Duhec, A. V., Jeanne, R. F., Maximenko, N., & Hafner, J. (2015). Composition and potential origin of marine debris stranded in the Western Indian Ocean on remote Alphonse Island, Seychelles. *Marine Pollution Bulletin, 96,* 76–86. https://doi.org/10.1016/j.marpolbul.2015.05.042

Dunlop, S. W., Dunlop, B. J., & Brown, M. (2020). Plastic pollution in paradise: Daily accumulation rates of marine litter on Cousine Island, Seychelles. *Marine Pollution Bulletin, 151,* 110803. https://doi.org/10.1016/j.marpolbul.2019.110803

Ebere, E. C., Wirnkor, V. A., Ngozi, V. E., & Chukwuemeka, I. S. (2019). Macrodebris and microplastics pollution in Nigeria: First report on abundance, distribution and composition. *Environmental Analysis, Health and Toxicology, 34,* e2019012. https://doi.org/10.5620/eaht.e2019012

Egressa, R., Nankabirwa, A., Ocaya, H., & Pabire, W. G. (2020). Microplastic pollution in surface water of Lake Victoria. *Science of the Total Environment, 741*, 140201. https://doi.org/10.1016/j.scitotenv.2020.140201

Eriksen, M., Lebreton, L. C. M., Carson, H. S., Thiel, M., Moore, C. J., Borerro, J. C., Galgani, F., et al. (2014). Plastic pollution in the world's oceans: More than 5 trillion plastic pieces weighing over 250,000 tons afloat at sea. *PLOS One, 9*, e111913. https://doi.org/10.1371/journal.pone.0111913

Evangeliou, N., Grythe, H., Klimont, Z., Heyes, C., Eckhardt, S., Lopez-Aparicio, S., et al. (2020). Atmospheric transport is a major pathway of microplastics to remote regions. *Nature Communications, 11*, 3381. https://doi.org/10.1038/s41467-020-17201-9

Fadare, O. O., & Okoffo, E. D. (2020). Covid-19 face masks: A potential source of microplastic fibers in the environment. *The Science of the Total Environment, 737*, 140279–140279. https://doi.org/10.1016/j.scitotenv.2020.140279

Fazey, F. M. C., & Ryan, P. G. (2016a). Biofouling on buoyant marine plastics: An experimental study into the effect of size on surface longevity. *Environmental Pollution, 210*, 354–360. https://doi.org/10.1016/j.envpol.2016.01.026

Fazey, F. M. C., & Ryan, P. G. (2016b). Debris size and buoyancy influence the dispersal distance of stranded litter. *Marine Pollution Bulletin, 110*, 371–377. https://doi.org/10.1016/j.marpolbul.2016.06.039

Fred-Ahmadu, O. H., Ayejuyo, O. O., & Benson, N. U. (2020). Microplastics distribution and characterization in epipsammic sediments of tropical Atlantic Ocean, Nigeria. *Regional Studies in Marine Science, 38*, 101365. https://doi.org/10.1016/j.rsma.2020.101365

Freeman, S., Booth, A. M., Sabbah, I., Tiller, R., Dierking, J., Klun, K., et al. (2020). Between source and sea: The role of wastewater treatment in reducing marine microplastics. *Journal of Environmental Management, 266*, 110642. https://doi.org/10.1016/j.jenvman.2020.110642

GESAMP. (1991). *The state of marine environment. Group of experts on the scientific aspects of marine pollution.* http://www.gesamp.org/publications/the-state-of-the-marine-environment

GESAMP. (2019). *Guidelines or the monitoring and assessment of plastic litter and microplastics in the ocean.* IMO/FAO/UNESCO-IOC/UNIDO/WMO/IAEA/UN/UNEP/UNDP/ISA Joint Group of Experts on the Scientific Aspects of Marine Environmental Protection. http://www.gesamp.org/publications/guidelines-for-the-monitoring-and-assessment-of-plastic-litter-in-the-ocean

GIZ. (2019). *Access to water and sanitation in Sub-Saharan Africa: Review of sector reforms and investments—Key findings to inform future support to sector development. Part 1. Synthesis report.* https://www.oecd.org/water/GIZ_2018_Access_Study_Part%20I_Synthesis_Report.pdf

González-Pleiter, M., Edo, C., Aguilera, Á., Viúdez-Moreiras, D., Pulido-Reyes, G., González-Toril, E., et al. (2021). Occurrence and transport of microplastics sampled within and above the planetary boundary layer. *Science of the Total Environment, 761*, 143213. https://doi.org/10.1016/j.scitotenv.2020.143213

Govender, J., Naidoo, T., Rajkaran, A., Cebekhulu, S., Bhugeloo, A., & Sershen. (2020). Towards characterising microplastic abundance, typology and retention in mangrove-dominated estuaries. *Water, 12*. https://doi.org/10.3390/w12102802

GRID-Arendal. (2005). *Vital water graphics: Major river basins of Africa.* https://www.grida.no/resources/5774

Guerranti, C., Martellini, T., Perra, G., Scopetani, C., Cincinelli, A. (2019). Microplastics in cosmetics: Environmental issues and needs for global bans. *Environmental Toxicology and Pharmacology, 68*, 75–79. https://doi.org/10.1016/j.etap.2019.03.007

Haddout, S., Gimiliani, G. T., Priya, K. L., Hoguane, A. M., Casila, J. C. C., & Ljubenkov, I. (2021). Microplastics in surface waters and sediments in the Sebou Estuary and Atlantic Coast, Morocco. *Analytical Letters*, 1–13. https://doi.org/10.1080/00032719.2021.1924767

Harris, P. T., Westerveld, L., Nyberg, B., Maes, T., Macmillan-Lawler, M., & Appelquist, L. R. (2021). Exposure of coastal environments to river-sourced plastic pollution. *Science of the Total Environment, 769*, 145222. https://doi.org/10.1016/j.scitotenv.2021.145222

Haseler, M., Schernewski, G., Balciunas, A., Sabaliauskaite, V. (2018). Monitoring methods for large micro- and meso-litter and applications at Baltic beaches. *The Journal of Coastal Conservation, 22*, 27–50. https://doi.org/10.1007/s11852-017-0497-5

Hoellein, T. J., Shogren, A. J., Tank, J. L., Risteca, P., & Kelly, J. J. (2019). Microplastic deposition velocity in streams follows patterns for naturally occurring allochthonous particles. *Scientific Reports, 9*, 3740. https://doi.org/10.1038/s41598-019-40126-3

Hoornweg, D., & Bhada-Tata, P. (2012). *What a waste: A global review of solid waste management.* World Bank. https://openknowledge.worldbank.org/handle/10986/17388

Horton, A. A., & Dixon, S. J. (2018). Microplastics: An introduction to environmental transport processes. *WIREs Water, 5*, e1268. https://doi.org/10.1002/wat2.1268

Hosoda, J., Ofosu-Anim, J., Sabi, E. B., Akita, L. G., Onwona-Agyeman, S., Yamashita, R., et al. (2014). Monitoring of organic micropollutants in Ghana by combination of pellet watch with sediment analysis: E-waste as a source of PCBs. *Marine Pollution Bulletin, 86*, 575–581. https://doi.org/10.1016/j.marpolbul.2014.06.008

Huntington, T. (2019). *Marine litter and aquaculture gear—White paper.* Report produced by Poseidon Aquatic Resources Management Ltd for the Aquaculture Stewardship Council. https://www.asc-aqua.org/wp-content/uploads/2019/11/ASC_Marine-Litter-and-Aquaculture-Gear-November-2019.pdf

Hurley, R., Woodward, J., & Rothwell, J. J. (2018). Microplastic contamination of river beds significantly reduced by catchment-wide flooding. *Nature Geoscience, 11*, 251–257. https://doi.org/10.1038/s41561-018-0080-1

Hwang, D.-J. (2020). The IMO action plan to address marine plastic litter from ships and its follow-up timeline. *Journal of International Maritime Safety, Environmental Affairs, and Shipping, 4*(2), 32–39. https://doi.org/10.1080/25725084.2020.1779428

Ilechukwu, I., Ndukwe, G., Mgbemena, N., & Akandu, A. (2019). Occurrence of microplastics in surface sediments of beaches in Lagos, Nigeria. *European Chemical Bulletin, 8*, 371. https://doi.org/10.17628/ecb.2019.8.371-375

IMO. (1972). *Convention on the prevention of marine pollution by dumping of wastes and other matter (London Convention).* IMO (International Maritime Organisation). https://www.imo.org/en/OurWork/Environment/Pages/London-Convention-Protocol.aspx

IMO. (1973/1978). *The international convention for the prevention of pollution from ships (marine pollution), 1973 as modified by the protocol of 1978.* https://www.imo.org/en/About/Conventions/Pages/International-Convention-for-the-Prevention-of-Pollution-from-Ships-(MARPOL).aspx

IMO. (2013). *Guide to good practice for port reception facility providers and users.* MEPC 671/Rev.1. http://www.mantamaritime.com/downloads/flag_news/MEPC.1_Circ.671_Rev.1.pdf

IMO. (2014). *Inadequacy of reception facilities: Revision of the IMO comprehensive manual on port reception facilities.* MEPC 67/11. http://www.basel.int/Portals/4/download.aspx?d=2014-MEPC-67-11.pdf

IMO. (2016). *Amendments to the Annex of the International Convention for the Prevention of Pollution from Ships, 1973, as modified by the protocol of 1978 relating thereto: Amendments to MARPOL Annex V. (HME substances and form of garbage record book).* Resolution Mepc.277(70). https://www.gard.no/Content/25062772/MEPC.277(70).pdf

International Pellet Watch. (2021). http://pelletwatch.org/

Iyare, P. U., Ouki, S. K., & Bond, T. (2020). Microplastics removal in wastewater treatment plants: A review. *Environmental Science: Water Research & Technology, 6*, 2664–2675. https://doi.org/10.1039/D0EW00397B

Jambeck, J., Hardesty, B. D., Brooks, A. L., Friend, T., Teleki, K., Fabres, J., et al. (2018). Challenges and emerging solutions to the land-based plastic waste issue in Africa. *Marine Policy, 96*, 256–263. https://doi.org/10.1016/j.marpol.2017.10.041

Jambeck, J. R., Geyer, R., Wilcox, C., Siegler, T. R., Perryman, M., Andrady, A., et al. (2015). Plastic waste inputs from land into the ocean. *Science, 347*, 768–771. https://doi.org/10.1126/science.1260352

Järlskog, I., Strömvall, A.-M., Magnusson, K., Gustafsson, M., Polukarova, M., Galfi, H., et al. (2020). Occurrence of tire and bitumen wear microplastics on urban streets and in sweepsand and washwater. *Science of the Total Environment, 729,* 138950. https://doi.org/10.1016/j.scitot env.2020.138950

Kanhai, L. D. K., Officer, R., Lyashevska, O., Thompson, R. C., & O'Connor, I. (2017). Microplastic abundance, distribution and composition along a latitudinal gradient in the Atlantic Ocean. *Marine Pollution Bulletin, 115,* 307–314. https://doi.org/10.1016/j.marpolbul.2016.12.025

Karlsson, T. M., Arneborg, L., Broström, G., Almroth, B. C., Gipperth, L., & Hasellöv, M. (2018). The unaccountability case of plastic pellet pollution. *Marine Pollution Bulletin, 129,* 52–60. https://doi.org/10.1016/j.marpolbul.2018.01.041

Katsumi, N., Kusube, T., Nagao, S., & Okochi, H. (2021). Accumulation of microcapsules derived from coated fertilizer in paddy fields. *Chemosphere, 267,* 10. https://doi.org/10.1016/j.chemos phere.2020.129185

Kay, P., Hiscoe, R., Moberley, I., Bajic, L., & Mckenna, N. (2018). Wastewater treatment plants as a source of microplastics in river catchments. *Environmental Science and Pollution Research, 25,* 20264–20267. https://doi.org/10.1007/s11356-018-2070-7

Kaza, S., Yao, L. C., Bhada-Tata, P., & Van Woerden, F. (2018). *What a waste 2.0: A global snapshot of solid waste management to 2050.* World Bank. https://openknowledge.worldbank.org/handle/ 10986/30317

Khatmullina, L., & Isachenko, I. (2017). Settling velocity of microplastic particles of regular shapes. *Marine Pollution Bulletin, 114,* 871–880. https://doi.org/10.1016/j.marpolbul.2016.11.024

Klöckner, P., Seiwert, B., Eisentraut, P., Braun, U., Reemtsma, T., & Wagner, S. (2020). Characterization of tire and road wear particles from road runoff indicates highly dynamic particle properties. *Water Research, 185,* 116262. https://doi.org/10.1016/j.watres.2020.116262

Knight, L. J., Parker-Jurd, F. N. F., Al-Sid-Cheikh, M., & Thompson, R. C. (2020). Tyre wear particles: An abundant yet widely unreported microplastic? *Environmental Science and Pollution Research, 27,* 18345–18354. https://doi.org/10.1007/s11356-020-08187-4

Knutsen, H., Cyvin, J. B., Totland, C., Lilleeng, Ø., Wade, E. J., Castro, V., et al. (2020). Microplastic accumulation by tube-dwelling, suspension feeding polychaetes from the sediment surface: A case study from the Norwegian Continental Shelf. *Marine Environmental Research, 161,* 105073. https://doi.org/10.1016/j.marenvres.2020.105073

Kole, P. J., Löhr, A. J., Van Belleghem, F. G. A. J., & Ragas, A. M. J. (2017). Wear and tear of tyres: A stealthy source of microplastics in the environment. *International Journal of Environmental Research and Public Health, 14,* 1265. https://doi.org/10.3390/ijerph14101265

Kooi, M., Nes, E. H. V., Scheffer, M., & Koelmans, A. A. (2017). Ups and downs in the ocean: Effects of biofouling on vertical transport of microplastics. *Environmental Science & Technology, 51,* 7963–7971. https://doi.org/10.1021/acs.est.6b04702

Krause, S., Baranov, V., Nel, H. A., Drummond, J. D., Kukkola, A., Hoellein, T., et al. (2021). Gathering at the top? Environmental controls of microplastic uptake and biomagnification in freshwater food webs. *Environmental Pollution, 268,* 115750. https://doi.org/10.1016/j.envpol. 2020.115750

Lamprecht, A. (2013). *The abundance, distribution and accumulation of plastic debris in Table Bay, Cape Town, South Africa.* University of Cape Town. Open UCT. https://open.uct.ac.za/han dle/11427/6633

Lane, S., Ahamada, S., Gonzalves, C., Lukambuzi, L., Ochiewo, J., Pereira, M., et al. (2007). *Regional overview and assessment of marine litter related activities in the west Indian Ocean region.* United Nations Environment Programme. https://wedocs.unep.org/bitstream/handle/20. 500.11822/8764/-A%20Regional%20Overview%20%26%20Assessment%20of%20Marine% 20Litter%20Related%20Activities%20in%20the%20%20Western%20Indian%20Ocean%20R egion-2007Regional_assessment_of_marine_litter_WIO_Region.PDF?sequence=3&isAllo wed=y

Lebreton, L., Slat, B., Ferrari, F., Sainte-Rose, B., Aitken, J., Marthouse, R., et al. (2018). Evidence that the Great Pacific garbage patch is rapidly accumulating plastic. *Scientific Reports, 8,* 4666. https://doi.org/10.1038/s41598-018-22939-w

Lebreton, L. C. M., Greer, S. D., & Borrero, J. C. (2012). Numerical modelling of floating debris in the world's oceans. *Marine Pollution Bulletin, 64,* 653–661. https://doi.org/10.1016/j.marpol bul.2011.10.027

Lebreton, L. C. M., Van Der Zwet, J., Damsteeg, J.-W., Slat, B., Andrady, A., & Reisser, J. (2017). River plastic emissions to the world's oceans. *Nature Communications, 8,* 15611. https://doi.org/10.1038/ncomms15611

Liu, F., Olesen, K. B., Borregaard, A. R., & Vollertsen, J. (2019). Microplastics in urban and highway stormwater retention ponds. *Science of the Total Environment, 671,* 992–1000. https://doi.org/10.1016/j.scitotenv.2019.03.416

Loulad, S., Houssa, R., Ouamari, N. E., & Rhinane, H. (2019). Quantity and spatial distribution of seafloor marine debris in the Moroccan Mediterranean Sea. *Marine Pollution Bulletin, 139,* 163–173. https://doi.org/10.1016/j.marpolbul.2018.12.036

Loulad, S., Houssa, R., Rhinane, H., Boumaaz, A., & Benazzouz, A. (2017). Spatial distribution of marine debris on the seafloor of Moroccan waters. *Marine Pollution Bulletin, 124,* 303–313. https://doi.org/10.1016/j.marpolbul.2017.07.022

Lourenço, P. M., Serra-Gonçalves, C., Ferreira, J. L., Catry, T., & Granadeiro, J. P. (2017). Plastic and other microfibers in sediments, macroinvertebrates and shorebirds from three intertidal wetlands of southern Europe and west Africa. *Environmental Pollution, 231,* 123–133. https://doi.org/10.1016/j.envpol.2017.07.103

Lusher, A., Hollman, P., & Mendoza, J. (2017). Microplastics in fisheries and aquaculture: Status of knowledge on their occurrence and implications for aquatic organisms and food safety. https://www.fao.org/3/I7677E/I7677E.pdf

Madzena, A., & Lasiak, T. (1997). Spatial and temporal variations in beach litter on the Transkei coast of South Africa. *Marine Pollution Bulletin, 34,* 900–907. https://doi.org/10.1016/S0025-326X(97)00052-0

Maes, T., Jessop, R., Wellner, N., Haupt, K., Mayes, A. G. (2017). A rapid-screening approach to detect and quantify microplastics based on fluorescent tagging with Nile Red. *Scientific Reports, 7,* 44501.

Maes, T., Barry, J., Leslie, H. A., Vethaak, A. D., Nicolaus, E. E. M., Law, R. J., et al. (2018). Below the surface: Twenty-five years of seafloor litter monitoring in coastal seas of North West Europe (1992–2017). *Science of the Total Environment, 630,* 790–798. https://doi.org/10.1016/j.scitotenv.2018.02.245

Mafuta, C., Barnes, R., Plummer, L., & Westerveld, L. (2018). *Sanitation and wastewater in Africa.* GRID-Arendal. https://www.grida.no/publications/471

Mahon, A. M., O'Connell, B., Healy, M. G., O'Connor, I., Officer, R., Nash, R., et al. (2017). Microplastics in sewage sludge: Effects of treatment. *Environmental Science & Technology, 51,* 810–818. https://doi.org/10.1021/acs.est.6b04048

Maione, C. (2021). Quantifying plastics waste accumulations on coastal tourism sites in Zanzibar, Tanzania. *Marine Pollution Bulletin, 168,* 112418. https://doi.org/10.1016/j.marpolbul.2021.112418

Marais, M., & Armitage, N. (2004). The measurement and reduction of urban litter entering stormwater drainage systems: Paper 2—Strategies for reducing the litter in the stormwater drainage systems. *Water SA, 30,* 483–492. https://doi.org/10.4314/wsa.v30i4.5100

Matsuguma, Y., Takada, H., Kumata, H., Kanke, H., Sakurai, S., Suzuki, T., et al. (2017). Microplastics in sediment cores from Asia and Africa as indicators of temporal trends in plastic pollution. *Archives of Environmental Contamination and Toxicology, 73,* 230–239. https://doi.org/10.1007/s00244-017-0414-9

Mayoma, B. S., Sørensen, C., Shashoua, Y., & Khan, F. R. (2020). Microplastics in beach sediments and cockles (*Anadara antiquata*) along the Tanzanian coastline. *Bulletin of Environmental Contamination and Toxicology, 105,* 513–521. https://doi.org/10.1007/s00128-020-02991-x

Migwi, F. K., Ogunah, J. A., & Kiratu, J. M. (2020). Occurrence and spatial distribution of microplastics in the surface waters of Lake Naivasha, Kenya. *Environmental Toxicology and Chemistry, 39*, 765–774. https://doi.org/10.1002/etc.4677

Mishra, S., Rath, C. C., & Das, A. P. (2019). Marine microfiber pollution: A review on present status and future challenges. *Marine Pollution Bulletin, 140*, 188–197. https://doi.org/10.1016/j.marpolbul.2019.01.039

Missawi, O., Bousserrhine, N., Belbekhouche, S., Zitouni, N., Alphonse, V., Boughattas, I., et al. (2020). Abundance and distribution of small microplastics (≤ 3 μm) in sediments and seaworms from the Southern Mediterranean coasts and characterisation of their potential harmful effects. *Environmental Pollution, 263*, 114634. https://doi.org/10.1016/j.envpol.2020.114634

Mobilik, J.-M., Ling, T.-Y., Husain, M.-L., & Hassan, R. (2016). Type and quantity of shipborne garbage at selected tropical beaches. *The Scientific World Journal, 2016*, 5126951. https://doi.org/10.1155/2016/5126951

Moss, K., Allen, D., González-Fernández, D., & Allen, S. (2021). Filling in the knowledge gap: Observing macroplastic litter in South Africa's rivers. *Marine Pollution Bulletin, 162*, 111876. https://doi.org/10.1016/j.marpolbul.2020.111876

Nachite, D., Maziane, F., Anfuso, G., & Williams, A. T. (2019). Spatial and temporal variations of litter at the Mediterranean beaches of Morocco mainly due to beach users. *Ocean & Coastal Management, 179*, 104846. https://doi.org/10.1016/j.ocecoaman.2019.104846

Naidoo, T., & Glassom, D. (2019). Sea-surface microplastic concentrations along the coastal shelf of KwaZulu-Natal, South Africa. *Marine Pollution Bulletin, 149*, 110514. https://doi.org/10.1016/j.marpolbul.2019.110514

Naidoo, T., Glassom, D., & Smit, A. J. (2015). Plastic pollution in five urban estuaries of KwaZulu-Natal, South Africa. *Marine Pollution Bulletin, 101*, 473–480. https://doi.org/10.1016/j.marpolbul.2015.09.044

Nel, H. A., Dalu, T., & Wasserman, R. J. (2018). Sinks and sources: Assessing microplastic abundance in river sediment and deposit feeders in an Austral temperate urban river system. *Science of the Total Environment, 612*, 950–956. https://doi.org/10.1016/j.scitotenv.2017.08.298

Nel, H. A., Dalu, T., Wasserman, R. J., & Hean, J. W. (2019). Colour and size influences plastic microbead underestimation, regardless of sediment grain size. *Science of the Total Environment, 655*, 567–570. https://doi.org/10.1016/j.scitotenv.2018.11.261

Nel, H. A., & Froneman, P. W. (2015). A quantitative analysis of microplastic pollution along the south-eastern coastline of South Africa. *Marine Pollution Bulletin, 101*, 274–279. https://doi.org/10.1016/j.marpolbul.2015.09.043

Nel, H. A., Hean, J. W., Noundou, X. S., & Froneman, P. W. (2017). Do microplastic loads reflect the population demographics along the southern African coastline? *Marine Pollution Bulletin, 115*, 115–119. https://doi.org/10.1016/j.marpolbul.2016.11.056

Nel, H. A., Naidoo, T., Akindele, E. O., Nhiwatiwa, T., Fadare, O. O., & Krause, S. (2021). Collaboration and infrastructure is needed to develop an African perspective on micro(nano)plastic pollution. *Environmental Research Letters, 16*, 021002. https://doi.org/10.1088/1748-9326/abdaeb

Nelms, S. E., Coombes, C., Foster, L. C., Galloway, T. S., Godley, B. J., Lindeque, P. K., Witt, M. J. (2017). Marine anthropogenic litter on British beaches: A 10-year nationwide assessment using citizen science data. *Science of the Total Environment, 579*, 1399–1409. https://doi.org/10.1016/j.scitotenv.2016.11.137

Nikiema, J., Figoli, A., Weissenbacher, N., Langergraber, G., Marrot, B., & Moulin, P. (2013). Wastewater treatment practices in Africa—Experiences from seven countries. *Sustainable Sanitation Practice, 12*, 26–34. http://www.ecosan.at/ssp/selected-contributions-from-the-1st-waterbiotech-conference-9-11-oct-2012-cairo-egypt/SSP-14_Jan2013_26-34.pdf

NOWPAP. (2009). *Port reception facilities in the NOWPAP region.* NOWPAP (Northwest Pacific Action Plan). https://www.unep.org/nowpap/resources/toolkits-manuals-and-guides/port-reception-facilities-nowpap-region-2009

Nunoo, F. K. E., & Quayson, E. (2003). Towards management of litter accumulation—Case study of two beaches in Accra, Ghana. *Journal of the Ghana Science Association, 5*, 145–155. https://www.scirp.org/(S(i43dyn45teexjx455qlt3d2q))/reference/ReferencesPapers.aspx?ReferenceID=1752341

Øhlenschlæger, J. P., Newman, S., & Farmer, A. (2013). *Reducing ship generated marine litter—Recommendations to improve the EU port reception facilities directive.* International Institute for European Environmental Studies. http://minisites.ieep.eu/assets/1257/IEEP_2013_Reducing_ship_generated_marine_litter_-_recommendations_to_improve_the_PRF_Directive.pdf

Okoffo, E. D., O'Brien, S., O'Brien, J. W., Tscharke, B. J., & Thomas, K. V. (2019). Wastewater treatment plants as a source of plastics in the environment: A review of occurrence, methods for identification, quantification and fate. *Environmental Science: Water Research & Technology, 5*, 1908–1931. https://doi.org/10.1039/C9EW00428A

Okoffo, E. D., Tscharke, B. J., O'Brien, J. W., O'Brien, S., Ribeiro, F., Burrows, S. D., Choi, P. M., Wang, X., Mueller, J. F., & Thomas, K. V. (2020). Release of plastics to australian land from biosolids end-use. *Environmental Science & Technology, 54*, 15132–15141. https://doi.org/10.1021/acs.est.0c05867

Okoffo, E. D., O'Brien, S., Ribeiro, F., Burrows, S. D., Toapanta, T., Rauert, C., et al. (2021). Plastic particles in soil: State of the knowledge on sources, occurrence and distribution, analytical methods and ecological impacts. *Environmental Science-Processes & Impacts, 23*, 240–274. https://doi.org/10.1039/D0EM00312C

Okuku, E., Kiteresi, L., Owato, G., Otieno, K., Mwalugha, C., Mbuche, M., et al. (2021a). The impacts of COVID-19 pandemic on marine litter pollution along the Kenyan Coast: A synthesis after 100 days following the first reported case in Kenya. *Marine Pollution Bulletin, 162*, 111840. https://doi.org/10.1016/j.marpolbul.2020.111840

Okuku, E. O., Kiteresi, L., Owato, G., Otieno, K., Omire, J., Kombo, M. M., et al. (2021b). Temporal trends of marine litter in a tropical recreational beach: A case study of Mkomani beach, Kenya. *Marine Pollution Bulletin, 167*, 112273. https://doi.org/10.1016/j.marpolbul.2021.112273

Okuku, E. O., Kiteresi, L. I., Owato, G., Mwalugha, C., Omire, J., Mbuche, M., et al. (2020a). Baseline meso-litter pollution in selected coastal beaches of Kenya: Where do we concentrate our intervention efforts? *Marine Pollution Bulletin, 158*, 111420. https://doi.org/10.1016/j.marpolbul.2020.111420

Okuku, E. O., Kiteresi, L. I., Owato, G., Mwalugha, C., Omire, J., Otieno, K., et al. (2020b). Marine macro-litter composition and distribution along the Kenyan Coast: The first-ever documented study. *Marine Pollution Bulletin, 159*, 111497. https://doi.org/10.1016/j.marpolbul.2020.111497

Oni, B. A., Ayeni, A. O., Agboola, O., Oguntade, T., & Obanla, O. (2020). Comparing microplastics contaminants in (dry and raining) seasons for Ox-Bow Lake in Yenagoa, Nigeria. *Ecotoxicology and Environmental Safety, 198*, 8. https://doi.org/10.1016/j.ecoenv.2020.110656

Onwuegbuchunam, D. E., Ebe, T. E., Okoroji, L. I., & Essien, A. E. (2017a). An analysis of ship-source marine pollution in Nigeria seaports. *Journal of Marine Science and Engineering, 5*, 39. https://doi.org/10.3390/jmse5030039

Onwuegbuchunam, D. E., Ogwude, I. C., Ibe, C. C., & Emenike, G. C. (2017b). Framework for management and control of marine pollution in Nigeria seaports. *American Journal of Traffic and Transportation Engineering, 2*, 59–66. https://doi.org/10.11648/j.ajtte.20170205.11

Osaloni, O. S. (2019). *The legal regulation of port waste management in the United Kingdom and Nigeria: Comparative analysis of Southampton Port in the UK and Apapa Port in Nigeria* [PhD]. The University of Central Lancashire. https://www.semanticscholar.org/paper/The-legal-regulation-of-port-waste-management-in-%3A-Osaloni/9878d4de4f6fe1423d7b93d9e6012a37977e6338

Pedrotti, M. L., Petit, S., Eyheraguibel, B., Kerros, M. E., Elineau, A., Ghiglione, J. F., et al. (2021). Pollution by anthropogenic microfibers in north-west Mediterranean Sea and efficiency of microfiber removal by a wastewater treatment plant. *Science of the Total Environment, 758*, 144195. https://doi.org/10.1016/j.scitotenv.2020.144195

Peters, J. K., & Marvis, A. N. (2019). *Assessment of port reception facilities and waste management control in Nigeria: Case study: (Tin Can Island Port)* [World Maritime University Dissertations]. https://commons.wmu.se/cgi/viewcontent.cgi?article=2213&context=all_dissertations

PlasticsSA. (2018). *All about plastics.*

Pohl, F., Eggenhuisen, J. T., Kane, I. A., & Clare, M. A. (2020). Transport and burial of microplastics in deep-marine sediments by turbidity currents. *Environmental Science & Technology, 54*, 4180–4189. https://doi.org/10.1021/acs.est.9b07527

Pramanik, B. K., Roychand, R., Monira, S., Bhuiyan, M., & Jegatheesan, V. (2020). Fate of road-dust associated microplastics and per- and polyfluorinated substances in stormwater. *Process Safety and Environmental Protection, 144*, 236–241. https://doi.org/10.1016/j.psep.2020.07.020

Preston-Whyte, F., Silburn, B., Meakins, B., Bakir, A., Pillay, K., Worship, M., et al. (2021). Meso- and microplastics monitoring in harbour environments: A case study for the Port of Durban, South Africa. *Marine Pollution Bulletin, 163*, 111948. https://doi.org/10.1016/j.marpolbul.2020.111948

Qi, Y. L., Beriot, N., Gort, G., Lwanga, E. H., Gooren, H., Yang, X. M., et al. (2020). Impact of plastic mulch film debris on soil physicochemical and hydrological properties. *Environmental Pollution, 266*, 9. https://doi.org/10.1016/j.envpol.2020.115097

REMPEC. (2005). *Port reception facilities: A summary of REMPEC's activities in the Mediterranean region.* REMPEC (Regional Marine Pollution Emergency Response Centre for the Mediterranean Sea). https://euroshore.com/sites/euroshore.com/files/downloads/port%20reception%20facilities%20final.pdf

Richardson, K., Hardesty, B. D., & Wilcox, C. (2019). Estimates of fishing gear loss rates at a global scale: A literature review and meta-analysis. *Fish and Fisheries, 20*, 1218–1231. https://doi.org/10.1111/faf.12407

Ryan, P. G. (1988). The characteristics and distribution of plastic particles at the sea-surface off the southwestern Cape Province, South Africa. *Marine Environmental Research, 25*, 249–273. https://doi.org/10.1016/0141-1136(88)90015-3

Ryan, P. G. (2014). Litter survey detects the South Atlantic 'garbage patch'. *Marine Pollution Bulletin, 79*, 220–224. https://doi.org/10.1016/j.marpolbul.2013.12.010

Ryan, P. G. (2020a). Land or sea? What bottles tell us about the origins of beach litter in Kenya. *Waste Management, 116*, 49–57. https://doi.org/10.1016/j.wasman.2020.07.044

Ryan, P. G. (2020b). The transport and fate of marine plastics in South Africa and adjacent oceans. *South African Journal of Science, 116*, 34–42.

Ryan, P. G., Bouwman, H., Moloney, C. L., Yuyama, M., & Takada, H. (2012). Long-term decreases in persistent organic pollutants in South African coastal waters detected from beached polyethylene pellets. *Marine Pollution Bulletin, 64*, 2756–2760. https://doi.org/10.1016/j.marpolbul.2012.09.013

Ryan, P. G., Dilley, B. J., Ronconi, R. A., & Connan, M. (2019). Rapid increase in Asian bottles in the South Atlantic Ocean indicates major debris inputs from ships. *Proceedings of the National Academy of Sciences, 116*, 20892–20897. https://doi.org/10.1073/pnas.1909816116

Ryan, P. G., Lamprecht, A., Swanepoel, D., & Moloney, C. L. (2014a). The effect of fine-scale sampling frequency on estimates of beach litter accumulation. *Marine Pollution Bulletin, 88*, 249–254. https://doi.org/10.1016/j.marpolbul.2014.08.036

Ryan, P. G., Maclean, K., & Weideman, E. A. (2020a). The impact of the COVID-19 lockdown on urban street litter in South Africa. *Environmental Processes, 7*, 1303–1312. https://doi.org/10.1007/s40710-020-00472-1

Ryan, P. G., & Moloney, C. L. (1990). Plastic and other artifacts on South-African beaches—Temporal trends in abundance and composition. *South African Journal of Science, 86*, 450–452. https://www.researchgate.net/publication/283507743_plastic_and_other_artefacts_on_South_African_beaches_temporal_trends_in_abundance_and_composition

Ryan, P. G., Moore, C. J., Van Franeker, J. A., & Moloney, C. L. (2009). Monitoring the abundance of plastic debris in the marine environment. *Philosophical Transactions of the Royal Society B-Biological Sciences, 364*, 1999–2012. https://doi.org/10.1098/rstb.2008.0207

Ryan, P. G., Musker, S., & Rink, A. (2014b). Low densities of drifting litter in the African sector of the Southern Ocean. *Marine Pollution Bulletin, 89*, 16–19. https://doi.org/10.1016/j.marpolbul. 2014.10.043

Ryan, P. G., & Perold, V. (2021). Limited dispersal of riverine litter onto nearby beaches during rainfall events. *Estuarine, Coastal and Shelf Science, 251*, 107186. https://doi.org/10.1016/j.ecss. 2021.107186

Ryan, P. G., Perold, V., Osborne, A., & Moloney, C. L. (2018). Consistent patterns of debris on South African beaches indicate that industrial pellets and other mesoplastic items mostly derive from local sources. *Environmental Pollution, 238*, 1008–1016. https://doi.org/10.1016/j.envpol. 2018.02.017

Ryan, P. G., Pichegru, L., Perold, V., & Moloney, C. L. (2020b). Monitoring marine plastics—Will we know if we are making a difference? *South African Journal of Science, 116*, 58–66. https:// sajs.co.za/article/view/7678/9944

Ryan, P. G., Weideman, E. A., Perold, V., Durholtz, D., & Fairweather, T. P. (2020c). A trawl survey of seafloor macrolitter on the South African continental shelf. *Marine Pollution Bulletin, 150*, 6. https://doi.org/10.1016/j.marpolbul.2019.110741

Ryan, P. G., Weideman, E. A., Perold, V., Hofmeyr, G., & Connan, M. (2021). Message in a bottle: Assessing the sources and origins of beach litter to tackle marine pollution. *Environmental Pollution, 288*, 117729. https://doi.org/10.1016/j.envpol.2021.117729

Ryan, P. G., Weideman, E. A., Perold, V., & Moloney, C. L. (2020d). Toward balancing the budget: Surface macro-plastics dominate the mass of particulate pollution stranded on beaches. *Frontiers in Marine Science, 7*, 14. https://doi.org/10.3389/fmars.2020.575395

Scheren, P. A., Ibe, A. C., Janssen, F. J., & Lemmens, A. M. (2002). Environmental pollution in the Gulf of Guinea—A regional approach. *Marine Pollution Bulletin, 44*, 633–641. https://doi.org/ 10.1016/s0025-326x(01)00305-8

Schmidt, C., Krauth, T., & Wagner, S. (2017). Export of plastic debris by rivers into the sea. *Environmental Science & Technology, 51*, 12246–12253. https://doi.org/10.1021/acs.est.7b02368

Schumann, E. H., Mackay, C. F., & Strydom, N. A. (2019). Nurdle drifters around South Africa as indicators of ocean structures and dispersion. *South African Journal of Science, 115*, 1–9. https:// doi.org/10.17159/sajs.2019/5372

Seeruttun, L. D., Raghbor, P., & Appadoo, C. (2021). First assessment of anthropogenic marine debris in mangrove forests of Mauritius, a small oceanic island. *Marine Pollution Bulletin, 164*, 112019. https://doi.org/10.1016/j.marpolbul.2021.112019

Shabaka, S. H., Ghobashy, M., & Marey, R. S. (2019). Identification of marine microplastics in Eastern Harbor, Mediterranean Coast of Egypt, using differential scanning calorimetry. *Marine Pollution Bulletin, 142*, 494–503. https://doi.org/10.1016/j.marpolbul.2019.03.062

Shruti, V. C., Pérez-Guevara, F., Elizalde-Martínez, I., & Kutralam-Muniasamy, G. (2020). Reusable masks for COVID-19: A missing piece of the microplastic problem during the global health crisis. *Marine Pollution Bulletin, 161*, 111777–111777. https://doi.org/10.1016/j.marpolbul.2020. 111777

Siegfried, M., Koelmans, A. A., Besseling, E., & Kroeze, C. (2017). Export of microplastics from land to sea. A modelling approach. *Water Research, 127*, 249–257. https://doi.org/10.1016/j.wat res.2017.10.011

Stoler, J. (2017). From curiosity to commodity: A review of the evolution of sachet drinking water in West Africa. *Wires Water, 4*, e1206. https://doi.org/10.1002/wat2.1206

Suaria, G., Achtypi, A., Perold, V., Lee, J. R., Pierucci, A., Bornman, T. G., et al. (2020a). Microfibers in oceanic surface waters: A global characterization. *Science Advances, 6*, eaay8493. https://doi. org/10.1126/sciadv.aay8493

Suaria, G., Perold, V., Lee, J. R., Lebouard, F., Aliani, S., & Ryan, P. G. (2020b). Floating macro- and microplastics around the Southern Ocean: Results from the Antarctic circumnavigation expedition. *Environment International, 136*, 105494. https://doi.org/10.1016/j.envint.2020. 105494

Taïbi, N.-E., Bentaallah, M. E. A., Alomar, C., Compa, M., & Deudero, S. (2021). Micro- and macro-plastics in beach sediment of the Algerian western coast: First data on distribution, characterization, and source. *Marine Pollution Bulletin, 165*, 112168. https://doi.org/10.1016/j.marpolbul.2021.112168

Talvitie, J., Mikola, A., Setälä, O., Heinonen, M., & Koistinen, A. (2017). How well is microlitter purified from wastewater?—A detailed study on the stepwise removal of microlitter in a tertiary level wastewater treatment plant. *Water Research, 109*, 164–172. https://doi.org/10.1016/j.watres.2016.11.046

Tata, T., Belabed, B. E., Bououdina, M., & Bellucci, S. (2020). Occurrence and characterization of surface sediment microplastics and litter from North African coasts of Mediterranean Sea: Preliminary research and first evidence. *Science of the Total Environment, 713*, 136664. https://doi.org/10.1016/j.scitotenv.2020.136664

Thiele, C. J., Hudson, M. D., Russell, A. E., Saluveer, M., & Sidaoui-Haddad, G. (2021). Microplastics in fish and fishmeal: An emerging environmental challenge? *Scientific Reports, 11*, 2045. https://doi.org/10.1038/s41598-021-81499-8

Thomas, M. L. H., Channon, A. A., Bain, R. E. S., Nyamai, M., & Wright, J. A. (2020). Household-reported availability of drinking water in Africa: A systematic review. *Water, 12*. https://doi.org/10.3390/w12092603

Torres, F. G., Dioses-Salinas, D. C., Pizarro-Ortega, C. I., & De-La-Torre, G. E. (2021). Sorption of chemical contaminants on degradable and non-degradable microplastics: Recent progress and research trends. *Science of the Total Environment, 757*, 143875. https://doi.org/10.1016/j.scitotenv.2020.143875

Toumi, H., Abidli, S., & Bejaoui, M. (2019). Microplastics in freshwater environment: The first evaluation in sediments from seven water streams surrounding the lagoon of Bizerte (Northern Tunisia). *Environmental Science and Pollution Research, 26*, 14673–14682. https://doi.org/10.1007/s11356-019-04695-0

Tramoy, R., Gasperi, J., Colasse, L., & Tassin, B. (2020). Transfer dynamic of macroplastics in estuaries—New insights from the Seine estuary: Part 1. Long term dynamic based on date-prints on stranded debris. *Marine Pollution Bulletin, 152*. https://doi.org/10.1016/j.marpolbul.2020.110894

Tsagbey, S. A., Mensah, A., & Nunoo, F. (2009). Influence of tourist pressure on beach litter and microbial quality—Case study of two beach resorts in Ghana. *West African Journal of Applied Ecology, 15*, 11–18. https://doi.org/10.4314/wajae.v15i1.49423

Uddin, S., Fowler, S. W., & Behbehani, M. (2020). An assessment of microplastic inputs into the aquatic environment from wastewater streams. *Marine Pollution Bulletin, 160*, 111538. https://doi.org/10.1016/j.marpolbul.2020.111538

UNEP. (1999). *Overview of land-based sources and activities affecting the marine, coastal and associated freshwater environment in the West and Central African Region* [UNEP Regional Seas Reports and Studies No. 171]. https://aquadocs.org/bitstream/handle/1834/800/UNEP171.pdf?sequence=1&isAllowed=y

UNEP. (2015). *Plastic in cosmetics*. https://wedocs.unep.org/handle/20.500.11822/21754

UNEP. (2018a). In P. Notten (Ed.). *Addressing marine plastics: A systemic approach—Stocktaking report*. United Nations Environment Programme. https://www.unep.org/resources/report/addressing-marine-plastics-systemic-approach-stocktaking-report

UNEP. (2018b). *Africa waste management outlook*. United Nations Environment Programme. https://wedocs.unep.org/handle/20.500.11822/25514

UNEP. (2019). *Small island developing states waste management outlook*. United Nations Environment Programme. https://www.unep.org/ietc/node/44

UNEP. (2020). *Monitoring plastics in rivers and lakes: Guidelines for the harmonization of methodologies*. https://wedocs.unep.org/bitstream/handle/20.500.11822/35405/MPRL.pdf

Van Calcar, C. J., & Van Emmerik, T. H. M. (2019). Abundance of plastic debris across European and Asian rivers. *Environmental Research Letters, 14*, 124051. https://doi.org/10.1088/1748-9326/ab5468

Van Der Mheen, M., Van Sebille, E., & Pattiaratchi, C. (2020). Beaching patterns of plastic debris along the Indian Ocean rim. *Ocean Science, 16*, 1317–1336. https://doi.org/10.5194/os-16-1317-2020

Van Dyck, I. P., Nunoo, F. K. E., & Lawson, E. T. (2016). An empirical assessment of marine debris, seawater quality and littering in Ghana. *Journal of Geoscience and Environment Protection, 04*, 21–36. https://doi.org/10.4236/gep.2016.45003

Van Sebille, E., Wilcox, C., Lebreton, L., Maximenko, N., Hardesty, B. D., Van Franeker, J. A., et al. (2015). A global inventory of small floating plastic debris. *Environmental Research Letters, 10*, 11. https://doi.org/10.1088/1748-9326/10/12/124006

Vapnek, J., & Williams, A. R. (2017). Regulating the packaged water industry in Africa: Challenges and recommendations, 20U. *Denver Water Law Review, 217*. https://repository.uchastings.edu/cgi/viewcontent.cgi?article=2631&context=faculty_scholarship

Velez, N., Zardi, G. I., Lo Savio, R., Mcquaid, C. D., Valbusa, U., Sabour, B., et al. (2019). A baseline assessment of beach macrolitter and microplastics along northeastern Atlantic shores. *Marine Pollution Bulletin, 149*, 110649. https://doi.org/10.1016/j.marpolbul.2019.110649

Verster, C., & Bouwman, H. (2020). Land-based sources and pathways of marine plastics in a South African context. *South African Journal of Science, 116*, 25–33. https://doi.org/10.17159/sajs.2020/7700

Vetrimurugan, E., Jonathan, M. P., Sarkar, S. K., Rodríguez-González, F., Roy, P. D., Velumani, S., et al. (2020). Occurrence, distribution and provenance of micro plastics: A large scale quantitative analysis of beach sediments from southeastern coast of South Africa. *Science of the Total Environment, 746*, 141103. https://doi.org/10.1016/j.scitotenv.2020.141103

Vilakati, B., Sivasankar, V., Mamba, B. B., Omine, K., & Msagati, T. A. M. (2020). Characterization of plastic micro particles in the Atlantic Ocean seashore of Cape Town, South Africa and mass spectrometry analysis of pyrolyzate products. *Environmental Pollution, 265*, 114859. https://doi.org/10.1016/j.envpol.2020.114859

Wakkaf, T., El Zrelli, R., Kedzierski, M., Balti, R., Shaiek, M., Mansour, L., et al. (2020). Characterization of microplastics in the surface waters of an urban lagoon (Bizerte lagoon, Southern Mediterranean Sea): Composition, density, distribution, and influence of environmental factors. *Marine Pollution Bulletin, 160*, 111625. https://doi.org/10.1016/j.marpolbul.2020.111625

Wallace, B., & Coe, J. M. (1998). *Guidelines for the provision of garbage reception facilities at ports under MARPOL Annex V*. NOAA Tech. Rep. U.S. Department of Commerce. https://aquadocs.org/handle/1834/20471?locale-attribute=en

Wang, H., Wang, T., Zhang, B., Li, F., Toure, B., Omosa, I. B., et al. (2014). Water and wastewater treatment in Africa—Current practices and challenges. *CLEAN—Soil, Air, Water, 42*, 1029–1035. https://doi.org/10.1002/clen.201300208

Weideman, E. A., Perold, V., Arnold, G., & Ryan, P. G. (2020a). Quantifying changes in litter loads in urban stormwater run-off from Cape Town, South Africa, over the last two decades. *Science of the Total Environment, 724*, 9. https://doi.org/10.1016/j.scitotenv.2020.138310

Weideman, E. A., Perold, V., Omardien, A., Smyth, L. K., & Ryan, P. G. (2020b). Quantifying temporal trends in anthropogenic litter in a rocky intertidal habitat. *Marine Pollution Bulletin, 160*, 10. https://doi.org/10.1016/j.marpolbul.2020.111543

Weideman, E. A., Perold, V., & Ryan, P. G. (2020c). Limited long-distance transport of plastic pollution by the Orange-Vaal River system, South Africa. *Science of the Total Environment, 727*, 11. https://doi.org/10.1016/j.scitotenv.2020.138653

Weideman, E. A., Perold, V., & Ryan, P. G. (2019). Little evidence that dams in the Orange-Vaal River system trap floating microplastics or microfibres. *Marine Pollution Bulletin, 149*, 110664. https://doi.org/10.1016/j.marpolbul.2019.110664

Weithmann, N., Moller, J. N., Loder, M. G. J., Piehl, S., Laforsch, C., & Freitag, R. (2018). Organic fertilizer as a vehicle for the entry of microplastic into the environment. *Science Advances, 4*, 7. https://doi.org/10.1126/sciadv.aap8060

Woodall, L. C., Robinson, L. F., Rogers, A. D., Narayanaswamy, B. E., & Paterson, G. L. J. (2015). Deep-sea litter: A comparison of seamounts, banks and a ridge in the Atlantic and Indian Oceans reveals both environmental and anthropogenic factors impact accumulation and composition. *Frontiers in Marine Science, 2*, 3. https://doi.org/10.3389/fmars.2015.00003

Wright, S. L., Ulke, J., Font, A., Chan, K. L. A., & Kelly, F. J. (2020). Atmospheric microplastic deposition in an urban environment and an evaluation of transport. *Environment International, 136*, 105411. https://doi.org/10.1016/j.envint.2019.105411

WWAP. (2015). *The United Nations world water development report 2015: Wastewater, the untapped resource.* United Nations World Water Assessment Programme. https://www.unwater.org/publications/world-water-development-report-2015/

WWAP. (2017). *The United Nations world water development report 2017: Wastewater, the untapped resource.* United Nations World Water Assessment Programme. https://wedocs.unep.org/handle/20.500.11822/20448

Yang, Z., Lü, F., Zhang, H., Wang, W., Shao, L., Ye, J., et al. (2021). Is incineration the terminator of plastics and microplastics? *Journal of Hazardous Materials, 401*, 123429. https://doi.org/10.1016/j.jhazmat.2020.123429

Zayen, A., Sayadi, S., Chevalier, C., Boukthir, M., Ben Ismail, S., & Tedetti, M. (2020). Microplastics in surface waters of the Gulf of Gabes, southern Mediterranean Sea: Distribution, composition and influence of hydrodynamics. *Estuarine, Coastal and Shelf Science, 242*, 106832. https://doi.org/10.1016/j.ecss.2020.106832

Zhang, Y., Kang, S., Allen, S., Allen, D., Gao, T., & Sillanpää, M. (2020). Atmospheric microplastics: A review on current status and perspectives. *Earth-Science Reviews, 203*, 103118. https://doi.org/10.1016/j.earscirev.2020.103118

Zhou, A., Zhang, Y., Xie, S., Chen, Y., Li, X., Wang, J., et al. (2020a). Microplastics and their potential effects on the aquaculture systems: A critical review. *Reviews in Aquaculture, 13*. https://doi.org/10.1111/raq.12496

Zhou, H., Zhou, L., & Ma, K. (2020b). Microfiber from textile dyeing and printing wastewater of a typical industrial park in China: Occurrence, removal and release. *Science of The Total Environment, 739*, 140329. https://doi.org/10.1016/j.scitotenv.2020.140329

Chapter 3
Impacts and Threats of Marine Litter in African Seas

Sumaiya Arabi, Yashvin Neehaul, and Conrad Sparks

Summary With a focus on plastic pollution, this chapter discusses the impacts of marine litter on the natural environment, the people and the economies of Africa. The impacts of marine litter will depend on various factors such as distribution, exposure time, size and type of organism. This chapter focusses on different impacts of marine litter at various scales, from ocean to coast, as well as more localised scales. The emphasis is on the coastal countries of the African continent, where information from Africa is lacking, and relevant data from other regions is used to infer possible impacts. Throughout this chapter, the environmental, social, economic and human impacts are discussed separately, although it should be remembered that these topics are intimately interlinked.

Keywords Environmental impacts · Economic · Social and human impacts · Waste management · Marine and coastal litter

3.1 Introduction

The first global accounts of plastic debris in the marine environment were reported in the 1970s (Carpenter & Smith, 1972; Carpenter et al., 1972; Cundell, 1974). One particular observation was made in 1971 during the 'Ra' Expedition (Heyerdahl, 1971) in the waters of Cape Verde, one of the African Small Island Developing States (SIDS). A brief history of marine litter research shows that since the 1960s concerns grew about the potential impacts of marine litter. From the first anecdotal reports of entanglement and plastic ingestion in the 1960s, scientific publications followed in

S. Arabi
Department of Forestry, Fisheries and the Environment, Cape Town, South Africa

Y. Neehaul (✉)
Reef Conservation, Les Flammants Branch Road Morcellement Pereyscape, Pereybere, Mauritius
e-mail: yashvin@chemist.com

C. Sparks
Department of Conservation and Marine Sciences, Cape Peninsula University of Technology, Cape Town, South Africa

© The Author(s) 2023
T. Maes and F. Preston-Whyte (eds.), *The African Marine Litter Outlook*,
https://doi.org/10.1007/978-3-031-08626-7_3

the 1970s, these were succeeded by a series of meetings on marine debris in the early 1980s which resulted by the end of the twentieth century in a better understanding of the marine litter issue and search for solutions (Ryan, 2015). Those early reports provided a first indication of the environmental catastrophe, which was in the making.

Scientific research on the impacts of plastic pollution is still ongoing, but the more we learn about the impacts of plastics, the gloomier the picture. The environmental impacts threaten the livelihoods of coastal populations through social, economic and human aspects. Throughout this chapter, the limited information on the impacts of marine pollution across Africa is highlighted. Even though the scientific interest in plastic pollution has increased over the years, the knowledge of the impacts on African countries is still largely unknown with most information restricted to South Africa (Fig. 3.1a–b). The scarceness of studies in Africa is indicative of limited funding across scientific fields, and the contribution of the African continent to global scientific knowledge was estimated at 2.8% in 2020 (Diop & Asongu, 2021). Interestingly, the Africa's contribution to the global GDP was also estimated at 2.8% (International Monetary Fund, 2021), showing that though limited, the studies are in line with what is available economically. It is noted, that even with knowledge gaps, enough is known about the impacts of marine litter and plastic pollution specifically, to implement mitigating actions and drive change ('Precautionary Principle'). Currently, most of the available data in Africa focusses on the presence, distribution and source determination of marine litter. This type of data provides a strong foundation to set baselines and determine impacts. Several global scientific initiatives such as capacity building, technology transfer and collaborations can contribute to promote marine plastic pollution research in Africa.

Box 3.1: The Special Case of Africa's Island States

During the United Nations Conference on Environment and Development held in Rio de Janeiro in 1992, SIDS were recognised as a discrete group of developing states facing specific social, economic and environmental vulnerabilities. These island states have limited land area, but they possess large exclusive economic zones at sea. Considering that a coastal population is commonly defined as the population residing within 100 km of the shoreline, the inhabitants of SIDS are exclusively coastal (Nicholls & Small,

2002; Small & Nicholls, 2003). The six African countries with SIDS status are Guinea-Bissau, Mauritius, Seychelles, Sao Tomé and Principé, Comoros and Cape Verde. Local economies are closely associated with the ocean as coastal tourism, fishing and related activities, aquaculture and more recently biotechnology are the main sectors included in the oceanic economy of SIDS.

Taking into account the remoteness and small, though dense, populations of these island states and their small contribution to marine plastic pollution, the impacts felt are uneven and disproportionate (Duhec et al., 2015; Onink et al., 2021). One particular example is Aldabra Atoll of the Seychelles. Designated as a UNESCO World Heritage site since 1982, this inhabited region harbours not only the largest population of giant tortoises in the world but also a wide variety of endemic animals. In March 2019, 25 tonnes of plastic litter were removed from the atoll at a cost of $224,537. This represented only 5% of the marine litter accumulated on the atoll. Removing the remaining litter would cost around $4.68 million and require 18,000 person-hours of labour (Burt et al., 2020).

3.2 Environmental Impacts of Marine Litter

Kühn et al. (2015) reviewed global publications on marine debris and reported that 557 species were affected by marine debris. The number of species reported to be affected by marine debris increased from 557 to 817 by 2016 (CBD, 2016). These studies showed that as with classification studies (refer to Chap. 2), plastic is the most encountered form of litter in the marine environment from an impact perspective.

Environmental impacts of marine litter are well known globally; however, information about the effects of marine litter in Africa is poor. Gall and Thompson (2015) categorised marine litter research per region, finding that the majority of studies ($n = 110$) were from North America, with only 12 impact studies from Africa. Marine litter is the cause of various negative environmental impacts globally, and these effects are arguably more pronounced in Africa due to the combination of poor waste management and rich biodiversity (see Chap. 1).

Akindele and Alimba (2021) reviewed 59 articles on the prevalence of plastic pollution from African aquatic environments in the period 1987–2020. Geographically, research outputs reported were as follows: 15 from North Africa (Algeria, Egypt, Morocco and Tunisia), six from East Africa (Ethiopia, Kenya, Tanzania and Uganda), 13 from West Africa (Ghana, Guinea-Bissau, Mauritania and Nigeria) and 25 studies from South Africa. The prevalence and effects of macro litter are the most prominent types of published research in Africa (Akindele & Alimba, 2021). Entanglement, smothering and ingestion by larger animals are well publicised due to the visible effects reported on marine mammals,

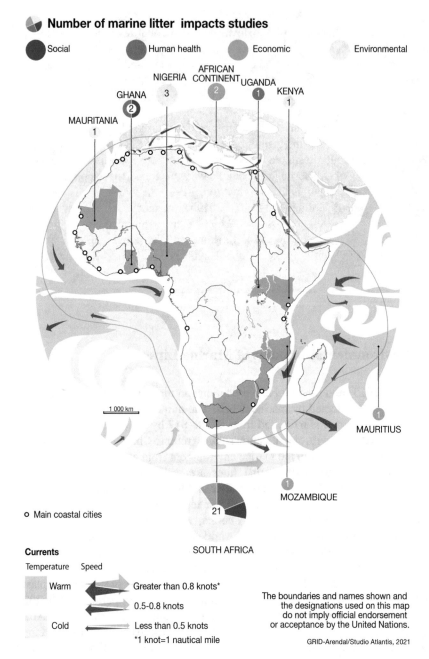

Fig. 3.1 **a** Total number of marine litter impact studies published across Africa in peer-reviewed journals (excluding quantification studies, which are covered in Chap. 2). *as of December 2021. **b** Total number of marine litter impact studies, by size fractionate, published across Africa in peer-reviewed journals (excluding quantification studies, which are covered in Chap. 2). *as of December 2021

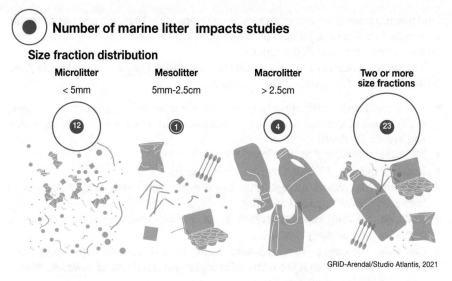

Fig. 3.1 (continued)

turtles and birds (Gregory, 2009; Ryan, 2018). The impacts of marine litter are often complex and the effects difficult to contextualise from micro to macroscale in terms of animals (cellular to biodiversity) and environments (localised to global). Although the following sections are compartmentalised, it should be noted that the impacts of marine litter in the environment are complex and interconnected.

3.2.1 Ingestion/Feeding

The ingestion of marine litter has been reported in over 519 species of animals (CBD, 2016), with records of publications increasing steadily (Ryan, 2015). Globally, ingestion of marine plastic litter has been recorded in at least 36% of seabird species (Ryan, 2018), 100% of turtle species, 59% of whale species and 36% of seal species (Kühn et al., 2015). Ingestion can be direct (primary ingestion) or indirect (secondary ingestion). Primary ingestion can be intentional or accidental. Intentional or deliberate ingestion of marine litter is when plastic items are mistaken for prey items and is influenced by foraging strategy, debris colour, age and sex of animals as well as characteristic of the litter (e.g. colour, size and chemical composition). Accidental ingestion occurs passively, mainly by non-selective feeders (e.g. filter feeders) (Kühn et al., 2015; Ryan, 2016). Secondary ingestion occurs by predators (and scavengers) consuming prey and food containing plastic items.

Ingestion studies tend to focus on the amount of plastic in the digestive tract of an organism. This amount is dependent on the ingestion rate and retention time (how long before removal via excretion and/or regurgitation). Thus, the amount of plastic in an organism will be dependent on the pollution level of the area the species forages in and its retention time (Ryan, 2016).

The main impacts of plastic ingestion (Derraik, 2002; Gregory, 2009; Mouat et al., 2010; Napper & Thompson, 2020) include:

- Accumulation of plastics in the digestive tract leading to damages such as wounds, scarring and ulceration, which in extreme cases can result in infection, starvation and eventually death.
- Mechanical blockage of the digestive tract.
- Reduced quality of life and reproductive capacity.
- Drowning, increased susceptibility to predators and death due to changes in buoyancy and/or impaired mobility.
- Reduced feeding capacity resulting in malnutrition, general debilitation, starvation and possibly death.
- Chemical poisoning from synthetic additives and contaminants comprising polymers that leads to reproductive disorders, increased risk of diseases, altered hormone levels and ultimately death.

Chemical effects from contaminants taken up through ingestion is dependent on equilibrium setting and thus retention times and partitioning coefficients. Chemical uptake is likely to be enhanced by longer retention times. Thus, *'species with broad, generalist diets that retain indigestible prey items in their digestive tracts for extended periods, probably are most likely to obtain large body burdens of hazardous chemicals from ingesting plastic items'* (Ryan, 2016). Further information round chemical impacts can be found under Sect. 2.5 Chemical Impacts.

A review of research on plastic ingestion in Africa between 1987 and 2020 found recorded ingestion in 63% of vertebrate species and 37% of invertebrate species studied (Akindele & Alimba, 2021). It is noted that this review excluded pre 1987 research. This meta-analysis of ingestion in Africa showed that plastic was found in 46% of examined fish species, 17% of birds species, 17% of molluscs species, 3% of plankton species and 7% of annelids species. Many of the species studied were reported as bioindicators of plastic ingestion or served as seafood across Africa (Akindele & Alimba, 2021). However, most of the research on plastic ingestion across Africa has been focused on fish species (Akindele & Alimba, 2021), most likely due to ease of access, as well as dependency on fish as a source of protein across the continent.

Most studies on plastic pollution across Africa come from South Africa (42% of reports) (Akindele & Alimba, 2021). Plastic ingestion by vertebrates in South Africa has been recorded in numerous species of birds ($n = 36$), sharks ($n = 10$), bony fish ($n = 7$) and turtles ($n = 1$) (Naidoo et al., 2020). Plastic ingestion by marine birds in South Africa is particularly well documented (Naidoo et al., 2020). Ryan (2008) reported that seabird ingestion of plastic particles consisted of mainly industrial pellets, but this may be changing, given the increase in fragmented plastics entering the environment (Ryan et al., 2020). The release of contaminants associated with plastic ingested by birds is important as these may be further contributing factors of the total impact of ingested plastics. Ryan et al. (2016) reported 60% of juvenile loggerhead turtles ($n = 24$) that died after stranding in the southern Cape of South Africa, contained ingested marine debris, of which 99% was plastic debris.

There is little information on ingestion of plastic by intertidal invertebrates that are not marine resources. One such example is Weideman et al. (2020a) who investigated the uptake of macroplastics by sea anemones (invertebrates) in southern Africa. These authors found that sandy anemones *Bunodactis reynaudi* in Cape Town, South Africa, often ingest plastic, mainly bags and other flexible packaging. These authors found that 491 litter items ingested by sandy anemones from 52 sampling events (9.4 ± 14.9 items month^{-1}) were mainly plastics, white in colour and correlated with high levels of beach litter items. Ingestion was more frequent during autumn, when the first winter rains had washed more litter into the sampling area. In addition to the field sampling, experiments indicated that sandy anemones *B. reynaudi* preferentially selected high-density polyethylene (HDPE) bags that were previously suspended in seawater for up to 20 days, suggesting that biofilms may enhance the potential for ingestion of plastic bags (Weideman et al., 2020a).

Microplastic ingestion is widespread across benthic and pelagic ecosystems where organisms feeding mechanisms do not allow for discrimination between prey and plastic items (Moore et al., 2001) or feed directly on microplastics, mistaking them for food (Moore, 2008). Microplastic ingestion research has increased over the past few years in Africa (see Table 3 in Alimi et al. (2021) and has been reported in freshwater birds (Reynolds & Ryan, 2018), fish (Bakir et al., 2020; Mbedzi et al., 2019; McGregor & Strydom, 2020; Naidoo et al., 2016; Shabaka et al., 2019; Sparks & Immelman, 2020), invertebrates such as zooplankton (Kosore et al., 2018), polychaetes (Nel & Froneman, 2018), mussels (Sparks, 2020; Wakkaf et al., 2020) and sea cucumbers (Iwalaye et al., 2020).

Research on marine and coastal microplastics in biota in Africa has been reported for marine resources (Abidli et al., 2018, 2019; Bakir et al., 2020; Sparks et al., 2021; Wakkaf et al., 2020). A study by Bakir et al., 2020 documented the levels of microplastics in three commercially important small pelagic fish species in South African waters, namely European anchovy (*Engraulis encrasicolus*), West Coast round herring (*Etrumeus whiteheadi*) and South African sardine (*Sardinops sagax*).

A higher concentration of microplastics for *S. sagax* (mean of 1.58 items individual^{-1}) compared to *Et. whiteheadi* (1.38 items individual^{-1}) and *En. encrasicolus* (1.13 items individual^{-1}) was found. The authors proposed *E. whiteheadi* as a bio-indicator for microplastics in South Africa.

Several studies that have shown that filter feeders, essentially shellfish, tend to accumulate microplastics in their gut (Karlsson et al., 2017; Lusher et al., 2017). Globally, coral polyps are known to have a particular taste for microplastic particles (Allen et al., 2017; Hall et al., 2015). Although most of the particles are rejected, 10–15% remain in the polyps. Additionally, Brown et al. (2008) showed that microplastics can even translocate to the circulatory system of mussels. The sorption of heavy metals, such as mercury, on the surface of microplastics is also of concern and can potentially contribute to the bioaccumulation of these toxic metals in shellfish, albeit this is dependent on equilibriums (Fernández et al., 2020).

Microplastics have been shown to impact invertebrates at a community level. Mussels in South Africa were able to produce more byssal threads when exposed to microplastic leachate seawater (when compared to a control), implying that mussel beds are influenced by plastic pollution (Seuront et al., 2021). An increased mortality in oysters who were chronically exposed to environmental relevant high loads of microplastics was observed in the laboratory. The results suggested that competitive abilities of intertidal bivalves may affect their ability to tolerate disturbance and ultimately influence their capacity as autogenic ecological engineers (Seuront et al., 2021). Marine animals are able to transfer ingested microplastics to predators when they occur in the natural environment. Maes et al., (2020a, b) found microplastics in North-East Atlantic porbeagle shark spiral valves, suggesting that these apex predators were consuming prey that had consumed microplastics. Southern mullet (*Chelon richardsonii*) sampled from a surf zone in South Africa recorded varied volumes of microplastics in guts from different ontogenetic stages (0–80 microplastic fibres across stages, 0–2 microplastic fragments across stages). This suggests that these fish are potential sources of microplastics (and associated contaminants) to be transferred up the food chain (McGregor & Strydom, 2020). Although microplastics are being reported at different trophic levels, the transfer and effects of contaminants associated with microplastics require further investigation in Africa's coastal ecosystems. Recently, the impacts of nanoplastics have been documented. They enter the marine organisms at the cellular level and have a wide range of impacts depending on the invaded organism (Piccardo et al., 2020). This is an emerging field of research in Africa and globally.

3.2.2 Entanglement

Entanglement in nets, ropes and other debris poses a significant risk to marine animals and has been recorded in 0.06% ($n = 92$) of invertebrate species such as corals (Schleyer & Tomalin, 2000), 0.27% ($n = 89$) of fish species, all 7 sea turtle species (Kühn et al., 2015), 36% of 414 seabird species (Ryan, 2018), 67% of 33 seal species and 31% of 80 marine mammal species worldwide (Kühn et al., 2015). It is important to note that entangled animals may be consumed by predators at sea or die and quickly sink, thereby eliminating them from potential detection in surveys (Gregory, 2009). Entanglement by marine litter is caused mostly by plastic items, in 91% of 205 species investigated for entanglement, 71% was due to plastic rope and netting (Gall & Thompson, 2015), and other specific items considered to be of high risk for entanglement of marine species are packing straps and six-pack rings (Ryan, 1990, 2018).

The main effects of entanglement (Akindele & Alimba, 2021; Derraik, 2002; Gall & Thompson, 2015; Gregory, 2009; Kühn et al., 2015; Laist, 1997; Mouat et al., 2010; Provencher et al., 2017; Sheavly & Register, 2007) include:

- Abrasions, cuts and wounds which can lead to infection, ulceration and ultimately death.
- Suffocation, strangulation and drowning of air-breathing species.
- Asphyxiation of species that require constant motion for respiration.
- Impaired mobility and reduced predator avoidance.
- Reduced fitness and increased energy cost of travel, due to entangled debris.
- Reduced ability to acquire food, which may ultimately lead to starvation.
- Restricted growth and prevention of circulation to limbs.
- Increased risk of sessile organisms being pulled off rocks by increased drag (e.g. corals, macroalgae, etc.).

Most research on entanglement in Africa has been reported in southern Africa. Naidoo et al. (2020) summarised marine plastic debris impacts in South Africa and reported plastic entanglement in sharks ($n = 8$ species), turtles ($n = 2$), mammals ($n = 5$) (Naidoo et al., 2020) and bird species ($n = 48$) (Ryan, 2018).

Ghost fishing refers to lost or abandoned fishing gear, including fish aggregating devices (FADs) (Balderson & Martin, 2015), which continues to entangle and ultimately kill organisms, as well as, destroy benthic habitats (Mouat et al., 2010). Ghost fishing affects an array of animals such as turtles, seabirds, seals and cetaceans, as well as commercially valuable and non-targeted fish species (Mouat et al., 2010; Stelfox et al., 2016). In addition to derelict fishing gear, other kinds of marine litter such as balloons, plastic bags and sheets are also known to cause entanglements (Kühn et al., 2015). It is also worth noting the difficulty in distinguishing between active and ghost gear at the time of entanglement, but the net effects are considered to be the same.

Anthropogenic factors relating to the mortality of 55 southern right whales (*Eubalaena australis*) off southern Africa between 1963 and 1998 indicated that five deaths were due to entanglement with active fishing gear (bycatch), with another 16 showing signs of non-fatal entanglements (Best et al., 2001). Between 1972 and 1979, Cape fur seals were reported to be affected by litter, specifically, fishing gear (nets, rope and lines), string, and plastic straps (Shaughnessy, 1980). Entanglement of fish (mainly sharks) in South Africa was mainly caused by plastic straps (from bait boxes and other packaging) in the 1980s (Ryan, 1990) and entanglement in shark nets (nets used for protection of bathers along beaches of KwaZulu Natal) (Cliff et al., 2002). Entanglement has also been recorded in other parts of (mainly northwest) Africa, which includes turtles (Duncan et al., 2017), seabirds (Rodríguez et al., 2013) and seals (Karamanlidis et al., 2008), with entanglement material often stemming from discarded fishing gear (Rodríguez et al., 2013).

3.2.3 Smothering

A large fraction of plastic marine litter tends to float in aquatic environments. As these litter items become heavier due to biofouling (Lobelle & Cunliffe, 2011), they have the potential to sink and settle on the seafloor (Fazey & Ryan, 2016), covering a variety of habitats from riverine, intertidal and near shore zones to abyssal environments (Gregory, 2009). The remaining plastics, that are denser the seawater, will sink and settle quicker, with the same impacts as their more buoyant counterparts. Plastic litter items settling on the seafloor may cause organisms to be smothered. This is of particular concern for marine vegetation and corals which also rely on light for primary production (Derraik, 2002; Kühn et al., 2015). Accumulation of litter may prevent gas exchange, resulting in reduced oxygen availability (Eich et al., 2015) and anoxic conditions in bottom waters, which themselves may be promoting climate-change conditions as a result of greater ocean stratification. The resulting impact on ecosystem functioning may be the covering of benthic organisms and changes in benthic ecosystem species composition and ecological interactions (Kühn et al., 2015; Napper & Thompson, 2020).

Although there are currently no reports on smothering caused by marine litter in Africa, Naidoo et al. (2020) reported that while South African coral reef diversity and associated sediments have been characterised, the susceptibility of these systems to marine debris was unclear.

3.2.4 Impact of Marine Litter Transport (Habitats and Dispersal)

The transport of fouling organisms and introduction of invasive species in habitat niches, such as in the African SIDS, has also been documented (Beaumont et al., 2019; Lachmann et al., 2017; Naidoo et al., 2020; Newman et al., 2015). The movement of flotsam is a natural occurrence, with wood, macroalgae and volcanic pumice being natural agents of flotsam dispersal for millions of years (Kiessling et al., 2015). Unlike natural flotsam, marine litter has no nutritive value (unless covered in a biofilm) and the additional amounts and features of litter (e.g. surface texture) are likely to influence colonisation and succession rates (Bravo et al., 2011). The 'plastisphere' is a term introduced by Zettler et al. (2013) to describe microbial communities on plastic marine debris. Plastics provide a substrate for proteins to develop biofilm formations that enable the debris to function as artificial 'microbial reefs' (Zettler et al., 2013). On entering the environment, biofilm and plastisphere development commences, further determining the pathway and fate of marine litter items.

Given the buoyant properties of many plastic items, oceanic and aquatic currents are able to transport plastic marine litter over vast areas (van Sebille et al., 2020). Most litter released from the coastal environment into the open ocean, if not settled on the benthos, eventually reaches beaches or remains afloat in the water column

(Onink et al., 2021). Depending on ocean current dynamics, marine litter has the potential to drift across entire oceans to other continental coastal areas (Ryan, 2020a), creating rafts which move alien species, pathogens, bacteria and hazardous substances including endocrine disruptors, persistent organic pollutants (POPs) and metals around the world (Naik et al., 2019). The durability of plastics also provides a platform to transport species from the sea surface, through the water column to ocean depths (Napper & Thompson, 2020). The movement of litter on the seafloor may also physically translocate benthic organisms (Naik et al., 2019). It is important to monitor floating litter, in terms of transport dynamics, estimation of fluxes of invasive species as well as assessment of the sources and pathways of litter in coastal areas. For example, Ryan (2020b) reported that plastic litter from local sources became less prominent with increased distance from urban areas in Kenya and South Africa (Ryan, 2020a), suggesting that localised sources of litter are major contributors to plastic pollution in urban coastal areas in the African sites sampled, with long-distance drift and transboundary transport being a varied concentration source across urban and remote areas. To develop a better understanding and to test policy interventions, a mass balance approach should be developed. Key information is missing for Africa and globally on plastic mass input, transfer and sink terms. The rates of accumulation, the dispersal pathways, the residence times in each compartment and the degradation rate into microplastics are unknown (Harris et al., 2021).

3.2.5   Chemical Impacts

Plastic contain additives, added during the production process, which can leach into the environment. High concentrations of chemical additives have the potential to be transferred from plastic litter to biota (Napper & Thompson, 2020; Rochman, 2015). Additionally, legacy pollutants including metals, POPs and endocrine disruptors (EDs) are sorbed onto plastic marine litter (Rochman, 2015). POPs have been reported in the marine and terrestrial environments and organisms in Africa (Alimi et al., 2020; Bruce-Vanderpuije et al., 2019). Ryan et al. (2012) showed plastic found in the marine and coastal environment to contain sorbed POPs. Hosoda et al. (2014) show more evidence of absorption of toxins, polychlorinated biphenyls (PCBs) from e-waste sorbed into plastics. Alimi et al. (2021) include a review of 14 studies of POPs and metals found in microplastics in the marine environment of Africa. Interestingly, Ryan (1988) found a correlation between the concentrations of PCBs in seabirds and the mass of ingested plastics, indicating that plastics can be a pathway for PCBs into organism tissues. More recently, Yamashita et al. (2021) identified flame retardants and legacy POPs in the preen gland oil of seabirds. The finding of these contaminants in blue petrals (*Halobaena caerulea*), who's range is limited to the remote region south of the Antarctic Polar Front is of particular interest.

The impact of POPs and metals on organisms when ingested is well studied, and their threats understood (as seen by the creation of the Stockholm Convention) (Mearns et al., 2018). Though it is acknowledged that the impact through ingestion of contaminated plastic is less studied, microplastics have been reported to adsorb POPs from its surrounding environment, and these POPs could be released following ingestion and/or be a pathway for transfer into tissues of animals (Galloway et al., 2017). The effect of plastics with regard to contaminant transfer is dependent on the context and linked to the setting of chemical equilibriums. In most cases, the net contribution of plastic ingestion to bioaccumulation of hydrophobic contaminants in marine biota is likely to be small in comparison with uptake of contaminants directly from water, sediment or food (Bakir et al., 2012; Koelmans et al., 2016). Ecotoxicological research on pollutants in marine debris has, however, shown that organic pollutants and metals have the potential to degrade the structure and function of ecosystems (Rochman, 2015). The impact of microplastics becomes evident at the onset of physiological processes being disrupted (subcellular protein function) causing diseases (Guzzetti et al., 2018), impaired activities such as reduced mobility and impaired reproduction (Sussarellu et al., 2016). The chemical impacts of plastic sorption and its pathways within the marine environment needs further research. The threat of absorption of contaminants from plastics to animals will depend on concentration and retention time, and bioaccumulation may have effects through the food chain. When considering plastics as a vector of contaminants, and their impacts, multiple sources and stress effects should be considered. It is imperative that such research is undertaken as contaminants sorbed to plastics have been shown to induce mutagenic or carcinogenic risks, endocrine disruption, genetic disruptions, inflammation, fibrosis and reproductive impairment (Arienzo et al., 2021). These effects are extrapolated to population, community and ecosystem levels and ultimately affect the productivity of entire ecosystems (Wright et al., 2013).

Box 3.2: Chemical Pollutants Found in the Marine Environment

As plastics can sorb and act as a vector for contaminates, concern arises for plastics to transport contaminants into different environmental compartments or remote areas, far from their sources, as well as provide a pathway, via ingestion, for bioaccumulation in species through bio-magnification and bio-concentration.

Litter of all sizes has been identified as a vector for toxic chemicals. For example, plastics are composed of the base monomer along with additives such as colourants, plasticisers, lubricants and flame retardants (Rochman et al., 2019). As the plastic materials are degraded into smaller plastic items in the environment, some of these residual monomers and chemicals are released into the aquatic system (Amelia et al., 2021; Dasgupta, 2021). Plastic particles in the oceans may sorb chemicals from the surrounding media (Näkki et al., 2021), and multi-stressor effect still need to be considered.

Persistent Organic Pollutants (POPs)

Plastics can sorb and act as a vector for POPs (Andrady, 2017; Ryan et al., 2012). POPs are highly toxic and are derived from diverse sources, including the combustion of some organic-bearing materials such as plastics and tyres that lead to the formation of 'unintentionally produced' furans, dioxins and polycyclic aromatic hydrocarbons. Some POPs have important industrial applications as pesticides, fire-retardants and as oil additives for electrical transformers. POPs undergo long-range transport in the environment and can easily reach the marine environment from land-based hotspots and diffuse sources, which include aerial deposition at sea. POPs can persist for decades in the environment and have been detected in coastal and marine environments of various sub-regions of Africa. Pesticide use in agricultural activities is believed to be the most likely source of POPs in southern Africa (UNEP/GPA, 2006).

In South Africa, Ryan et al. (2012) used PE pellets obtained from three beaches to monitor the concentrations of POPs over two decades and observed that there was a trend towards decreasing concentrations. In Lagos, Nigeria, phthalate esters were found to have been absorbed onto microplastics collected from littoral sandflat sediments at five beaches and three lagoon locations (Benson & Fred-Ahmadu, 2020). Total phthalate esters concentrations ranged from 0 to 164 mg kg^{-1} dry weight, dominated by di(2-ethylhexyl) phthalate (DEHP), dibutyl phthalate (DnBP) and dimethyl phthalate. It was suggested that future studies of POPs in total sediment versus the microplastics fraction might be useful for refining ecological risk assessments. Similarly, at eleven different beaches of the Ghanaian coastline, plastic resin pellets were found to contain PCBs (Hosoda et al., 2014). PCB concentrations (13 congeners) were higher in beaches off Accra and Tema (39–69 ng g^{-1}-pellets) than those in smaller coastal towns (1–15 ng g^{-1}-pellets) which are close to global backgrounds, indicating local inputs of PCBs near urban centres. Mansour (2009) reported various POPs in waters and sediments of the Nile River and some lakes close to the coastal zones of Egypt since the early 1980s. Several studies have also been conducted in Nigeria in which environmental media were shown to be contaminated with POPs (Adeyemi et al., 2019, Williams and Mesubi, 2013). Pesticide use in agricultural activities is believed to be the most likely source of POPs in southern Africa (UNEP/GPA, 2006). Most African countries are parties to the Stockholm Convention, the international treaty that seeks to eliminate the global scourge of POPs in the environment (Chap. 4). Adherence to the principles of the convention will assist African countries to ultimately and significantly reduce their burdens of toxic POPs.

Most African countries are parties to the Stockholm Convention, the international treaty that seeks to eliminate the global scourge of POPs in the environment (Chap. 4). Adherence to the principles of the convention will assist African countries to ultimately and significantly reduce their burdens of toxic POPs.

Heavy Metals

Hazardous metals have been detected in both the marine environment and marine organisms. For example, at several locations off the coasts of Cameroon in central Africa, marine sediments showed enrichment of arsenic, cadmium, cobalt, chromium, copper, iron, manganese, nickel, lead, vanadium and zinc (Biney et al., 1994). Similar findings have been reported in the literature for Côte d'Ivoire (Affian et al., 2009), Nigeria (Bamanga et al., 2019), Morocco (Maanan, 2008) and South Africa (Orr et al., 2008). Plastics can also sorb and act as a vector for metals (Naik et al., 2019). Metals have been observed in microplastics in Nigeria (Fred-Ahmadu et al. 2020).

In South Africa, mercury contamination of the marine environment has also been reported by Walters et al. (2011), while a long-term (1985–2007) dataset on heavy metals (copper, cadmium, iron, lead, mercury, zinc and manganese) in the marine environment is available from the International Mussel Watch Programme. The mussel watch data indicates that metal concentrations in *Mytilus galloprovincialis* showed no detectable increase over the study period (Sparks et al., 2014).

Petroleum Hydrocarbon

Lastly, plastic absorbs oil from seawater (Aboul-Gheit et al., 2006). Petroleum hydrocarbon oil pollution of the marine environment occurs due to releases from coastal and offshore oil exploration and production activities, as well as accidental and deliberate spillages which occur from ships that traverse the busy African and international waterways.

Oil spillage is particularly common around the coasts of African countries that are major oil producers such as Nigeria, Angola and Gabon (UNEP, 2013, 2021). The East African route is also characterised by heavy use of oil tankers. Oil spills are particularly detrimental to marine ecosystem quality, rendering the water largely unusable for aquaculture, recreation and transportation, and killing many large and small organisms within a short period. In some cases, seafood is tainted with smell of the petroleum hydrocarbons, making seafood unusable for human consumption. The Niger Delta and nearby coastal regions are particularly well known for the environmental degradation and security crisis that has been caused to the areas by oil spills to land and water (Kadafa, 2012). Recent global statistics have revealed that oil spill incidents of varied magnitudes are known to have occurred around and off most coastal seaports in the African continent (ITOPF, 2020). More recently, in July 2020, about

1000 tonnes of oil was accidentally spilled off the coast of Mauritius when the cargo ship MV Wakashio ran aground on a coral reef on the southeast tip of the country and smeared about 1.5 km stretch of the coastline (Lewis, 2020).

Marine litter is considered to be an emerging marine contaminant, especially given increased knowledge about chemicals associated with microplastics. The known impacts are centred around entanglement, ingestion and subsequent physical (Wright et al., 2013) and toxicological effects on biota (Browne et al., 2013). Larger microplastics (0.1–5 mm) may impact digestive systems (Lusher, 2015), while smaller, nano-sized particles are able to permeate lipid membranes of invertebrates, resulting in deformed membrane structure and ultimately cellular dysfunction (Alimba et al., 2021; Rossi et al., 2014).

3.2.6 Climate Change and Ecological Impacts

Plastics in the environment are contributing to the climate change, with current greenhouse emissions from the plastics industry estimating to contribute to a global temperature increase of 1.5 °C by 2050 (Hamilton et al., 2019). Additionally, plastics act as threat multipliers to climate change (UNEP, 2021); for example, the plastic pollution acts as an insulator increasing the temperature of beaches, which in addition to increasing global temperatures can affect the biodiversity of the beaches (Lavers et al., 2021; Sevwandi Dharmadasa et al., 2021). In an effort to mimic the effects of climate change on microplastics uptake, sea cucumbers sampled from KwaZulu-Natal, South Africa, were fed polyethylene fragments at different concentrations and at different temperatures. Ingestion rates increased with higher microplastic concentrations and temperatures up to 28 °C (ingestion rates decreased at temperatures >28 °C). More microplastics were also retained at 28 °C, with these results suggesting that the effects of microplastics on biota will become more pronounced with increasing temperature related to climate change (Iwalaye et al., 2021). Similar to the social, economic and human tragedies associated with climate change, the environmental impacts of marine plastic pollution directly affect the livelihoods of coastal populations.

Ecological impacts of litter are complex. Although open waste disposal or dump sites provide migratory and resident birds with nesting and feeding sites, there are risks of birds ingesting plastics and becoming entangled in litter. This changes the ecology of the species involved, specifically natural ecological activities pertaining to foraging and reproduction in natural habitats (Reusch et al., 2020). The exact effect on an ecological level is unknown, as supplemental feeding will have a positive effect on survival rates, which may offset entanglement and ingestion effects. Given the poor waste management across Africa (Willis et al., 2018), it is probable that the prevalence of large amounts of plastic litter may be far reaching, across the entire continent.

Changes in biodiversity, from entanglement and ingestion of marine litter, may have implications for survival of endangered species (CBD, 2016; Gall & Thompson, 2015). In some cases, the changes in landscape, food and ecological interactions, due to marine litter, may result in an increase in biodiversity. For example, due to litter aggregating in marine benthic regions, new habitats become available where organisms settle on plastic items (Song et al., 2021; Weideman et al., 2020b).

Box 3.3: COVID-related impacts

The COVID-19 pandemic has resulted in an increase in the use of personal protective equipment (PPE) for both citizens and frontline workers (e.g. face masks, face screens, gloves, portable hand sanitizer and full protective clothing). Due to poor waste management practices, an increase in PPE has been observed in the environment globally and in Africa (Okuku et al., 2021; Ryan et al., 2020). The increased observation of littered PPE (Okuku et al., 2021; Ryan et al., 2020; Thiel et al., 2021) is detailed in Chap. 2.

PPE such as masks is comprised of polymers such as polypropylene and/or polyethylene, polyurethane, polystyrene (Ammendolia et al., 2021; Fadare & Okoffo, 2020; Selvaranjan et al., 2021) and gloves comprised of PVC, latex and nitrile (De-la-Torre & Aragaw, 2021). Once the PPE ends up in the coastal environment, these degrade and contribute to microplastics contamination (Fadare & Okoffo, 2020).

In coastal organisms, the presence of PPE can cause impacts due to entanglement, ingestion and smothering—though depending on numbers this impact of PPE specifically may be trivial compared to overall marine litter. The monitoring of effects and mitigation measures of PPE is limited both globally and across Africa.

3.3   Social, Economic and Human Impacts

The interaction between humans and the ocean is important for our social, economic and mental well-being. Humans rely on the marine environment for food sources both from a subsistence and economic perspective (refer to Chap. 1). The ocean also plays an important role in terms of recreation, shipping and tourism (Newman et al., 2015), see Chap. 1 for more details. The presence of marine litter can impact these activities, as well as have a potential negative impact on human health (Van der Meulen et al., 2014). The social and human health impacts of marine litter are not well understood worldwide, even less so in Africa. There is a lack of published literature that explores the impacts of marine litter on the economic and social well-being of humans and its effects on human health. This section focuses on what is known about the social, economic and human impacts of marine litter in Africa. Most literature of social impacts of marine litter focuses on South Africa, albeit still with many gaps in knowledge. The remaining African countries had little, to no, literature available.

3.3.1 Social Impacts

The social impacts of marine litter consider its effects on the quality of peoples' lives, which can include: the loss of non-use values, impacts on cultural services, recreation and aesthetics (Ballance et al., 2000; Mouat et al., 2010). The interconnected social and economic impact on safety and navigation is also discussed below.

Loss of Non-Use Value and Cultural Services

Non-use value relates to the positive impact on a person in knowing that an ecosystem, species or resource exists, and that it will be around for future generations. This value is unaffected by whether or not the person visits the place (Mouat et al., 2010). Studies have found that visiting the ocean can have positive impacts on people's mood and can even reduce an individual's blood pressure (UN Environment, 2017). However, the presence of marine litter on a beach can result in negative mood changes (Arabi & Nahman, 2020; Beaumont et al., 2019; GESAMP, 2015; UNEP, 2016).

The marine environment contributes towards emotional and/or cultural services. People can feel attached and attracted to animals such as dolphins, whales and turtles. They also form part of cultural heritage to some groups. The potential loss of these animals can have an impact on peoples' well-being (Beaumont et al., 2019; UNEP & GRID-Arendal, 2016).

The marine environment contributes to spiritual and/or religious services. Many religions identify the interface between land and sea as a place where they can receive intercession with their deity (Preston-Whyte, 2008). In the African SIDS particularly, there is a strong spiritual link to the sea. Indeed, the ocean represents both freedom from oppression and a memorial for all the lives lost at sea during transportation and exploitation in colonial times (Baderoon, 2009). There is ancient symbolism in the cleansing during immersion that takes place during religious beach ceremonies, for example: in South Africa, black South Africans in Durban have a strong cultural connection with the beach (Preston-Whyte, 2008). It is common to experience the sound of drums together with singing which announces the pre-dawn ceremony. Worshippers pray and sing and are dowsed in the waves as part of ceremonial rituals (Preston-Whyte, 2008).

No studies were found in Africa that show specifically the impacts of marine litter on non-use or cultural values.

Reduced Recreational Activities and Aesthetic Value

Beaches and oceans are used for a variety of recreational activities such as swimming, diving, paddle boarding, scuba diving, kitesurfing and wave surfing. Surfing is a recreational sport but can also be considered having a cultural value to many communities, defining their way of life (Booth, 2005). The presence of marine litter can have negative impacts on recreational users from both an aesthetic and safety perspective (Beaumont et al., 2019).

> **Box 3.4: Case Study—Marine Litter Impact on Tourism:**
>
> A study by Balance et al. (2000) in Cape Town, South Africa, interviewed local and non-local beach users to determine the perceived importance of beach cleanliness. Foreign tourists in particular rated beach cleanliness as the number one factor in choosing a beach to visit. Approximately half of the people interviewed stated that they were willing to spend more than seven times an average trip cost to visit a clean beach (it is noted that this is a stated preference, not an actual measured response). In addition, 44% of residents were willing to travel 50 km or more to visit a clean beach. The presence of more than 10 large litter items per meter of beach would deter 97% of visitors from visiting that beach again, reducing the recreational value by R300,000 per year. The total impact on the regional economy could equate to a loss of billions of Rands per year. The estimated total annual recreational value of specific beaches in the Cape Peninsula was at least R3–23 million.

The presence of marine litter on a beach results in a loss of aesthetic value, which impacts the way people enjoy the environment (Beaumont et al., 2019; Werner et al., 2016), thus affecting a person's quality of life. Not visiting the coast due to the presence of litter can also have physical, mental and emotional health implications due to reduced physical activity and the lack of social interactions with family and friends (Arabi & Nahman, 2020; Beaumont et al., 2019). A study in Accra, Ghana, found that residents are concerned about unclean beaches. Unclean beaches were one of the top issues identified by the participants of the study (Van Dyck et al., 2016). People in Accra seem to be desensitised to years of litter campaigns and consider authorities to be responsible for the issue of marine litter (Van Dyck et al., 2016). People tend to litter more in areas that are already littered (Van Dyck et al., 2016). The negative impacts on the aesthetics of a beach due to the presence of litter have been seen along the Benin coasts and Port Bouet, Vridi, Grand Bassam in Côte d'Ivoire, which are popular tourist beaches (UNEP, 1999). Sharp objects in marine litter can also cause health issues by injuring beach users and can discourage local people from using the beach for recreational activities such as playing football or exercising.

From an education and social change perspective, a study in Durban, South Africa found that beach goers had a negative perception towards single-use plastics and had a high understanding of the impacts of single-use plastics on the environment. These beachgoers stated that they were willing to reduce their use of single-use plastics to dampen these environmental impacts (Van Rensburg et al., 2020).

Safety and Navigational Hazards

Litter, particularly plastics tend to clog drains, waterways and sewers when there is heavy rainfall. This results in damage to properties, weakening of infrastructure and can be of risk to lives (Sambyal, 2018; Turpie et al., 2019). In 2018, clogged drains during a heavy rainfall event in Accra, Ghana, resulted in the loss of 150 lives (Sambyal, 2018). In Malawi, flooding has become a common occurrence where, in 2019, flash flooding in the City of Lilongwe damaged 179 households and possibly two lives were lost due to the clogging of drains by plastic litter (Turpie et al., 2019). The building up of litter in drains and rivers and the subsequent flushing of high volumes of litter during rainfall events (as observed in Biermann et al., 2020) is also likely to increase navigation issues. These problems need to be addressed by improving waste management, urban planning and draining maintenance. However, this is costly and most African countries cannot afford to implement such activities (Turpie et al., 2019). The Emergency Services Department in Malawi clears drains monthly or on an *ad-hoc* basis. The collected waste is left on the side of the road because the Department does not have the resources to transport the waste for disposal so the waste re-enters the environment resulting in a continuous cycle (Turpie et al., 2019).

The presence of marine litter in ocean waters, particularly discarded fishing gear, ropes and plastic bags present hazards to navigation of vessels and personnel. From other regions, we know that propellers get tangled by discarded ropes and fishing lines resulting in vessel instability, plastic bags clog water intakes resulting in damage to pumps and collision with litter can result in damage to propellers which can cause injuries to personnel or even death (Mouat et al., 2010; Newman et al., 2015; Ten Brink et al., 2009; Turpie et al., 2019; UNEP, 2016).

Impacts on safety and navigation have both a social and economic impact. In 2010, the estimated cost of repairs and lost time at sea, from fishing gear and other macroplastic blocking inlet pipes and entangling propellers, was approximately 5% of total fishing revenue (Mouat et al., 2010).

Marine litter can delay the response time of emergency services in cases of rescue missions at sea due to the entanglement of propellers and clogging of inlet valves of rescue vessels. This delay in responding to a rescue mission can result in death of personnel that may have required urgent medical assistance (Abalansa et al., 2020).

Marine litter can delay the response time of emergency services in cases of rescue missions at sea due to the entanglement of propellers and clogging of inlet valves of rescue vessels. This delay in responding to a rescue mission can result in death of personnel that may have required urgent medical assistance (Abalansa et al., 2020).

The impacts of marine litter have been seen on occasion in the Port of Durban, South Africa, after heavy rainfall events. These events also result in considerable unplanned cleaning costs for Ports and municipalities. Estimated clean-up costs of marine litter due to storm events in the Port of Durban in 2019 ranged between ZAR52 800 (USD3 400) and ZAR1 046000 (USD68 400) and totalled ZAR4 350000 (USD284 800) during that period alone (Arabi & Nahman, 2020). The costs related to the impacts of marine litter can be quite significant and have longer term impacts on the economics of a country.

3.3.2   Economic Impacts of Marine Litter

The economic impacts of marine litter consider its negative effects of a monetary nature, specifically affecting safety and navigation (discussed in Sect. 3.3.1.3, Safety and navigational hazards), fisheries, cultural services and ecosystem services. In 2005, the World Resources Institute published the Millennium Ecosystem Assessment (MEA), which provided a framework that categorises ecosystem services into provisioning services, supporting services, regulatory services and cultural services (Ecosystems & Human Well-being, 2005). Each of these categories includes several economic actors (e.g. fisheries and aquaculture) that are directly impacted by marine plastic pollution (Haines-Young & Potschin, 2012). The MEA categorisation of ecosystem services is used to discuss the impacts on fisheries, cultural services and ecosystem services below. The economies of plastic stakeholders are also discussed below. The interconnected social and economic impact on safety and navigation is discussed above under social impacts.

The Economies of Marine Litter

Environmental economics considers marine litter to be a 'public bad' which is both non-excludable and non-rivalrous (Common & Stagl, 2005). As is the case for most environmental 'public bad', marine plastic pollution is an example of market failure that can be attributed to both a missing market and negative externalities (Common & Stagl, 2005; Oosterhuis et al., 2015). The missing market results from the absence of a definition of an acceptable level of marine litter from those involved in the production of plastic and those requesting a reduction in marine plastic litter. Finding agreement between the two groups is not an easy task considering the large number of individuals, private companies, organisations and governments involved. In the middle, we have the consumers, who contribute, both directly and indirectly, to marine litter. The involvement of waste managers and recyclers is also essential in these discussions. Such negotiations, although difficult to put in place, can result in the conception of suitable schemes and initiate the setting up of a circular economy. The Ellen MacArthur Foundation initiated such discussions internationally in 2017 and the Global Commitment 2020 Progress Report attests to the resultant benefits (Ellen MacArthur Foundation, 2017, 2020). Nevertheless, with the exception of South Africa, African states are not yet involved in this initiative. For further discussions on the circular economy, see Chap. 4. The market failure in the marine plastic litter problem is also ascribed to the existence of negative externalities. These are usually described as adverse side effects of the actions of the producers and the consumers that impact the welfare or production of others. For example, the fisheries and the tourism industry are adversely affected by marine litter (Beaumont et al., 2019; UNEP, 1999; Viool et al., 2019). Since the costs of the undesired side effects are not incurred by the producers and those involved in the act of marine littering, there is no financial incentive to promote a behavioural change on individual, industrial or institutional scale.

The plastic manufacturing industry is flourishing due to the high demand for plastic products and the low price of the plastic raw materials. Babayemi et al. (2019) reported that from 1990 to 2017, 117.6 Mt of plastics entered the African continent through 33 countries: 86.1 Mt of primary plastics (pellets) and 31.5 Mt as final products. This figure excludes local production, for industrialised countries like South Africa, where local production outstrips importation. Six countries with significant contributions to this imported amount were Egypt (18.4%), Nigeria (16.9%), South Africa (11.6%), Algeria (11.2%), Morocco (9.6%) and Tunisia (6.9%). By 2030, the continent is predicted to consume 344 Mt of plastic (Babayemi et al., 2019). To put this figure into perspective, in 2019, almost 370 Mt were produced globally (Plastic Europe, 2019). The setting up of a circular economic strategy is a sustainable and necessary solution, but it involves significant financial investment. Dedicated and location specific life cycle analysis are required to determine most cost effective, humane and environmental options. In some cases, alternatives to plastics can be more costly and may have less efficient physical properties or other adverse environmental effects. In a linear economy, focusing on financial benefits of the producers, manufacturing plastic remains a better strategy. When taking into account the impacts of marine plastic pollution on ecosystem services and their associated values, the cost for the production of new virgin plastic would rise significantly, and a circular economy would become more attractive.

Impacts on Provisioning Services: Fisheries and Aquaculture

The fisheries sector is probably one of the economic actors that is the most impacted by marine litter, while paradoxically being a major contributor to the problem (Arabi & Nahman, 2020). It is estimated that abandoned, lost, discarded fishing gear (ALDFG) from industrial and artisanal fishing makes up 48% of the mass of plastic in the infamous North Pacific gyre Lebreton et al. (2018) and less than 10% by volume of the plastics found in the ocean overall (Macfadyen et al., 2009). Commonly known as 'ghost gear', they continue to catch fish years after they have been lost in the ocean. Consequently, ghost fishing contributes to the ongoing depletion of fish stocks. Several studies indicate that ghost fishing decreases landed catches of market species by 0.5–30% in various regions and competes effectively against fishers for their daily catch (Brown & Macfadyen, 2007; Laist, 1987; Sancho et al., 2003; Santos et al., 2003; Sukhsangchan et al., 2020).

Although ALDFG is often blamed for decreasing fish stocks in Europe and North America, scientific data is lacking for the African continent (Gilman et al., 2016). East of Africa, Al-Masroori et al. (2004) investigated the catch rate of simulated lost fish traps near Muscat and Mutrah, Sultanate of Oman. The study estimated that ghost fishing mortality was 1.34 kg/trap/day. An exponential model was used to evaluate the total mass of fish caught over different time periods, and it predicted that each trap would catch 67 and 78 kg during 3 and 6 months, respectively (Al-Masroori et al., 2004). More recently, Randall (2020) summarised the potential impacts of ALDFG on the South African fisheries sector by extrapolating the Global Ghost Gear Initiative (GGGI) methodology to estimate the impacts of several gear classes. The report suggests that the fishing sector with the greatest risk of ALDFG is the gillnet sector, the second highest risk is in the sectors of West Coast rock lobster (trap only, not hoopnet), South Coast rock lobster and the exploratory octopus trap fishery. The remaining fisheries have a low risk of creating ALDFG (Randall, 2020). Richardson et al. (2019) provide a baseline estimate that can be extrapolated to Africa, i.e. 5.7% of all fishing nets, 8.6% of all traps and 29% of all lines were lost to the world's oceans in 2017 (Richardson et al., 2019). The absence of data does not imply that ghost fishing is not affecting African countries. On the contrary, a reduction in fish catch on the African continent can potentially have severe repercussions on the availability of food. An initiative is underway by the Sustainable Seas Trust, through the African Marine Waste Network, to estimate the socioeconomic costs of ALDFG in African seas (Sustainable Seas Trust, 2021).

The cost of navigational interference by ALDFG and other litter is covered in Sect. 3.3.2.1. Safety and navigational hazards. Furthermore, the cleaning and repair of fishing gear with trapped plastic debris is among the additional activities that most industrial fishing companies have to consider in their operations (Macfadyen et al., 2009).

Another economic sector that is both affected and contributes to the ocean plastic problem is caged aquaculture, particularly the shellfish farming industry. Typically, the farming structures are made of metal wires coated with PVC or other equivalent plastics, to protect them from rusting. Furthermore, a considerable number of polypropylene ropes is used for mooring purposes. Over time, these ropes wear and photodegrade, breaking down into microplastics that can be easily ingested by wild and farmed marine organisms. Though there is global literature on the contribution of aquaculture to marine litter, there is an absence of data on how the plastic material is managed during and after its usage or affect the organisms in and outside these commercial farms. In general, there is an absence of scientific studies on the distribution and impacts of marine litter, and plastics from sea-based sources and the African continent is no exception (Gilardi et al., 2020). The aquaculture industry is relatively small in Africa, but growing (see details in Chap. 1), and so its contribution to marine litter, and marine litters effects on it are expected to increase.

The impacts of plastic litter on individual organisms, as discussed in the Sect. 3.2 Environmental impacts of marine litter, are well documented, but translating these impacts to fish stocks is not an easy task. Considering that several other stress factors, such as over-fishing and climate change, also contribute significantly to marine fish stock depletion, the distinct impact of marine litter is difficult to assess and can often be considered a threat multiplier (UNEP, 2021), rather than a standalone stressor.

A practical fishing technique that is growing worldwide among needy communities is the use of mosquito nets as fishing gear. This practice is used in at least 15 African countries (Short et al., 2018). Mosquito nets are either cheap or free in countries affected by malaria and are used as beach seines or drag nets. However, the small mesh size (0.6–1.2 mm) catches juvenile fishes and contributes to fish stock depletion (Jones & Unsworth, 2020). As they are not built for fishing purposes, they often break while in operation. The effect of these nets and anti-mosquito chemicals they often carry as marine litter is unknown.

In 2016, fisheries and aquaculture directly contributed 1.3% to the African GDP and employed over 12 million people (58% in the fishing and 42% in the processing sector) (Tall et al., 2016). Employment multiplier effects are remarkable in certain regions: for example, for every fishers' job, 1.04 additional onshore jobs are created in Mauritania, and this ratio increases to 3.15 in Guinea (de Graaf & Garibaldi, 2014). From an economic perspective, these ratios demonstrate the potential for further job creation through value chain development in the African fisheries and aquaculture sector. Therefore, considering the existing global pressures on the fisheries sector, a reduced daily catch as a result of marine litter should by all means be prevented to protect the regional economic drivers of this sector in the coming years.

Economic Impacts on Cultural Services: Recreation, Aesthetics and Heritage

Another economic sector that is directly impacted by marine litter is the tourism industry. Pre-COVID, tourism contributed on average 9–10% of the GDP of SIDS (World Bank Group, 2015). For example, tourism contributed 24.4% of the GDP of the Seychelles in 2013 (World Bank Group, 2013) compared to 7% in continental Africa.

The SIDS tourism relies on beautiful, clean, sandy beaches, yet these island states are the most disproportionately impacted by marine litter (van der Mheen et al., 2020). With small land areas and relatively small human populations, the consumption of plastics is proportionally modest when compared to continental states. However, their large exclusive economic zones harbour a considerable amount of plastic originating from the most polluting states (Lachmann et al., 2017). The Seychelles is a good example. Computational models generated from sea surface currents and windage, as well as empirical evidence from brand audits, have shown that the majority of plastic debris accumulating on their beaches originate from South East Asia (Duhec et al., 2015; Dunlop et al., 2020).

The tourism industry responds by continuous cleaning of targeted beaches at an additional cost. However, remote areas are left to accumulate large litter loads (Burt et al., 2020). Quite often, voluntary clean-up commitments by NGOs or government work programmes (such as Working for the Coast Programme, South Africa) are the sole clean-up campaigns for these regions (Ryan & Swanepoel, 1996). In the current global COVID pandemic, travel restrictions and temporary closure of borders are common. The resulting impacts on the tourism industry are severe, with considerable loss of revenue (Škare et al., 2021). Several hotels are on technical temporary closure pending a return to normal (Chummun & Mathithibane, 2020). Their beaches are currently not being cleaned.

Impact on Ecosystem Services

As described thoroughly in the previous sections (see Sect. 3.2.4), marine organisms are interacting with this unprecedented presence and abundance of plastic. A recent laboratory study on four globally distributed coral species indicates that ingested microplastics are encrusted in the calcium carbonate structure (Hierl et al., 2021). The long-term effects on these reef-building organisms are not known, but corals may become an unexpected microplastic sink. Considering that coral reefs are already under enormous pressure caused by global phenomena such as ocean acidification and climate change, the ever-growing amount of plastic in the marine environment will worsen the strain on corals. The threat of marine litter to ecosystem services, from a global perspective, is captured in Chap. 1. Data for Africa on this topic is lacking; however, marine litter is considered a threat multiplier for coastal ecosystems (UNEP, 2021).

Box 3.5: IOC Study

Aware of the threats posed by unmanaged plastic waste, the Indian Ocean Commission (IOC) has initiated programmes for its member countries, which include three African SIDS (Mauritius, Seychelles and Comoros), as well as La Réunion and Madagascar. In 2014, an initial study evaluated and mapped waste management systems in the region (Fig. 3.2). The resulting report included several recommendations to optimise waste management in the IOC member states (Indian Ocean Comission, 2021a). One noteworthy recommendation is the need to address waste management at the regional level. In 2019, the IOC published an ambitious regional action plan to specifically enhance the regional plastic waste management system and also to pave the way for the Expédition Plastique dans l'Océan Indien (ExPLOI) project that is expected to start in 2021 (Indian Ocean Comission, 2021b).

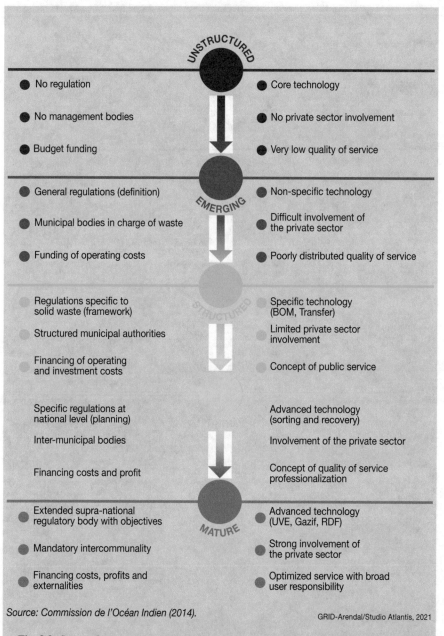

Source: Commission de l'Océan Indien (2014).

GRID-Arendal/Studio Atlantis, 2021

Fig. 3.2 Status of waste management in the IOC member countries (2014–2019)

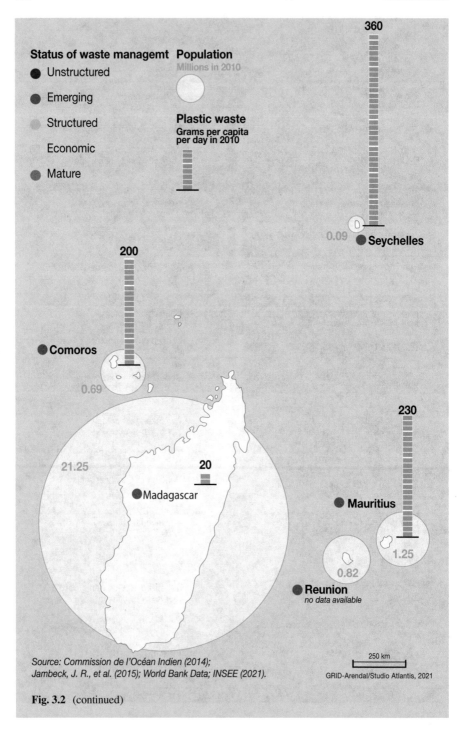

Fig. 3.2 (continued)

Financed by the Agence Française du Developpement, this ambitious endeavour intends to address the different aspect of plastic waste and pollution management in the region. Through the SWIOFISH2 programme, the IOC is extending the regional initiative to the AIODIS (Indian Ocean and African Island Developing States) (Indian Ocean Comission, 2021c). This unique platform of eight countries that include all the African SIDS (Cape Verde, Guinea-Bissau, Sao Tome and Principe, Comoros, Mauritius, Madagascar, Maldives and Seychelles) is an opportunity to collaborate, to share experiences and meet specific challenges such as improving the sustainable management of their vast maritime territories, developing their Blue Economies and promoting circular economies. Marine plastic pollution is on the priority list of this collaboration. Even if the data is cruelly scarce on plastic pollution and the related impacts in the African island states, the ongoing collaborative projects and initiatives will contribute to narrowing the gap in the near future.

3.3.3 Human Health Impacts

Considering the reliance of subsistence fishing as a food source particularly in Africa, the potential impacts of marine litter on human health are concerning. Human health impacts could be direct as a result of injuries and death, as well as indirect, e.g. ecosystem decline, loss of nutrition, chemical and other risks. Discarded containers have been shown to influence the seasonal distribution of dengue mosquitoes in rural settings in India (Shukla et al., 2020). Focused research is needed to understand the extent of this risk. Once we understand the risks, mitigation measures can be put in place to educate and advise communities of the impacts of marine litter on their health. It is worth noting that the World Health Organization (WHO) regards microplastics as a minor human health issue at this time (Naidoo et al., 2020; WHO, 2019).

Transfer Through the Food Chain

Microplastics present in the marine environment are ingested by marine organisms. When organisms are consumed as a whole, this forms a direct pathway to humans through the food web, thereby potentially affecting human health. Microplastics have been found in fish and shellfish which are commonly consumed by humans. It is of particular concern in shellfish, oysters, mussels, sea urchins, sea cucumbers and small fish which tend to be eaten whole without removal of the digestive tract (Arabi & Nahman, 2020; Landrigan et al., 2020; Turpie et al., 2019; UNEP, 2016;

Werner et al., 2016). In Tunisia, it is estimated that consumption of local mussels results in the ingestion of an estimated 4.2 microplastics capita^{-1}, year^{-1} (Wakkaf et al., 2020). For South Africa, human consumption of microplastics by mussels was estimated to be 3.03 microplastics capita^{-1} year^{-1} (Sparks et al., 2021). Three commercially important small pelagic fish species in South African waters, namely European anchovy (*E. encrasicolus*), West Coast round herring (*E. whiteheadi*) and South African sardine (*S. sagax*), were found to contain on average at least 1 microplastic per fish (Bakir et al., 2020). It should be noted that microplastics can also be ingested via other food sources, including honey, beer and tap and bottled water. Due to their small size, microplastics can also be inhaled, similar to fine particulate matter (Chen et al., 2020; De-la-Torre, 2020); however, no studies exist on these topics in Africa yet.

To date, research looking at the incidences of endocrine disruption and the ingestion of plastics are largely lacking. Although certainly possible, there is currently only limited evidence to support it (Amereh et al., 2019; Chen et al., 2021; Rochman et al., 2014). Guttered fish are still an area of potential concern. The consumption of dried fish is popular in South Africa and dates back to the seventeenth century. Although these are gutted, and therefore microplastics in the gut may be removed, there is still the potential for chemical accumulation in other tissues (Naidoo et al., 2020). In addition, the drying process requires a large amount of salt which has also been found to be contaminated with microplastics (Naidoo et al., 2020).

Endocrine disruption has been associated with chemical additives used in the plastics industry such as bisphenol A (BPA), phthalates and brominated flame retardants. Endocrine disrupting chemicals can affect the unborn foetus, children at early developmental stages and adolescents, as well as the general population. These can have human health impacts if introduced into the human body either for medical purposes or through accidental inhalation or ingestion (Arabi & Nahman, 2020; Godswill & Gospel, 2019; Turpie et al., 2019; UNEP, 2016). Studies have also looked at the ability of plastics to sorb environmental pollutants such as heavy metals, POPs, including polychlorinated biphenyl (PCBs), polybrominated diphenyl ethers (PBDEs), organochlorine pesticides (OPCs) such as dichlorobiphenyl trichloroethane (DDTs), hexachlorocyclohexanes (HCHs), polycyclic aromatic hydrocarbon (PAHs) alkylphenols, bisphenol A (BPA), parabens, estrogenic steroids and metals (cadmium, aluminium and zinc) on their surfaces (Menéndez-Pedriza & Jaumot, 2020; Scutariu et al., 2019). Newer unregulated compounds replacing previously identified toxic chemicals are also a concern. In addition, there is concern of marine plastics interacting with pharmaceuticals such as antibiotics, antidepressants and beta-blockers (Menéndez-Pedriza & Jaumot, 2020; Scutariu et al., 2019). The extent of the impacts of these on organisms, including humans, is not understood. Plastic pollutants have been found in over 83% of tap water samples around the world. This study suggested that individuals could be consuming 3000–4000 plastic particles from tap water annually (Godswill & Gospel, 2019), as such ingestion of plastic through seafood needs to be considered in line with other ingestion pathways as well as the sorption potential of chemicals in those pathways.

The large amounts of marine debris in the ocean have resulted in a substrate for microbial colonisation and a new potential route of dispersal, thereby supporting microbial communities (Werner et al., 2016), including antibiotic-resistance bacteria (Moore et al., 2020). This causes concerns regarding the transport of pathogens on marine litter and its possible impact on environmental and human health aspects (Naidoo et al., 2020; Turpie et al., 2019; UNEP, 2016; Werner et al., 2016). The need for further research on this is imperative to fully understand the scale of the problem which could have possible implications for the aquaculture sector and the Blue Economy in Africa. See Chap. 1 for details on the Blue Economy in Africa.

Spreading of Diseases

As discussed earlier, litter clogs drains and storm water which could lead to flooding during periods of high rainfall. The plastic containers and hollow surfaces can hold water themselves, increasing the risk of mosquito breeding grounds and therefore increasing the risks of malaria. There is also some evidence of marine litter supporting cholera and bacteria that cause gastrointestinal diseases (Krystosik et al., 2020; Newman et al., 2015; UNEP, 2016). In Kampala, Uganda, flooding led to five cholera outbreaks between a period of 11 years. Increased risk of disease outbreak due to a lack of proper waste disposal has been found in Malawi during the wet season. In 2018, 929 cholera cases were recorded which resulted in 30 deaths (Turpie et al., 2019). *Aedes aegypti* is a species of mosquito that breeds in stagnant water in artificial substrates such as discarded tyres, cans and plastic containers and has been linked to the spread of the Zika virus. In 2007, a Zika virus outbreak occurred in West Africa and spread into subtropics. The spread of such a virus is exacerbated by poor waste collection and management (UNEP, 2016). A study conducted in Dar es Salaam City, Tanzania, in 2014 during a dengue fever virus outbreak found that the most common breeding grounds for *Aedes* mosquitoes were discarded plastic containers and tyres (Mboera et al., 2016).

An increase in sea level, wind speed, wave height and altered rainfall conditions will lead to an increased amount of floating plastic debris along coastal areas. These increased amounts of plastic debris can result in negative health impacts for recreational ocean users (Keswani et al., 2016). Plastic debris, microplastic particles and fibres in the marine environment can transport hazardous microorganisms, including vectors for human disease (Keswani et al., 2016; Landrigan et al., 2020). In a study in Zanzibar, plastic litter from four rural sites was analysed for bacteria. Diverse bacterial species, of which many were multidrug resistant, were found on the plastic waste items including three human pathogens: *Citrobacter freundii*, *Klebsiella pneumoniae* and *Vibrio cholerae*. Plastics were therefore confirmed to act as reservoirs for bacterial growth which can lead to the transmission of infectious diseases and antimicrobial resistance (Rasool et al.,

2021). *Escherichia coli* and other pathogenic species have been detected on plastics in the marine environment and on public beaches resulting in the exposure of humans to these pathogens (Keswani et al., 2016; Landrigan et al., 2020). There is a need for more focused research in this area to identify all the risks related to marine litter on beaches and the ocean with regards to the spreading of diseases.

Hazards to Swimmers, Divers and Waste Pickers (Cuts, Abrasions and Needle Injuries)

Beach users are at risk of injury due to the presence of broken glass, pieces of metal, sharp plastic fragments and medical waste often found in marine litter. A risk that is often not considered is the exposure during clean-ups, or by individuals (e.g. beach combers, waste pickers and homeless individuals) who in countries such as Sierra Leone sort through marine litter containing broken glass and sharp objects such as needles (Sankoh, 2021, personal communications). Some of these communities do not have the necessary protective equipment such as masks and gloves when sorting through waste and are therefore more exposed to possible injury or to pathogens which can lead to respiratory infections, skin diseases, chronic diseases and mental illness. They often lack the knowledge of the impacts of exposure to waste on their health (Made et al., 2020). A study in Johannesburg, South Africa, found that waste pickers at dumpsites tend not to visit clinics for medical help and assessments due to the fear of being judged or discriminated against (Made et al., 2020).

Discarded fishing nets and ropes can cause risk to swimmers and divers who can get tangled in them (Beaumont et al., 2019; Tsagbey et al., 2009; Werner et al., 2016). In a study in Accra, Ghana, representatives from four different environmental organisations experienced injuries such as wounds, diseases and discomfort from marine litter on beaches (Van Dyck et al., 2016).

Leaching of Poisonous Chemicals

Components of plastics like plasticizers and additives can be toxic to human health due to the leaching of chemicals. The amount of toxic chemicals in the ocean is relatively low, but this can become important when large amounts of debris with high levels of toxic compounds are accidentally deposited into the ocean (Werner et al., 2016), such as during the M/V X-Press Pearl nurdle spill. Exposure to combustion, heat and chemicals led to agglomeration, fragmentation, charring and chemical modification of the plastic, creating an unprecedented complex spill of visibly burnt plastic and unburnt nurdles. This added chemical complexity included combustion-derived polycyclic aromatic hydrocarbons. A portion of the burnt material contained petroleum-derived biomarkers, indicating that it encountered some fossil-fuel products during the spill (de Vos et al., 2021).

3.4 Conclusions

Most of the African data available on marine plastic litter focuses on the distribution and sources (refer to Fig. 2.1a–b, Chap. 2). This provides a strong foundation and an optimistic outlook for the coming years in understanding the impacts of marine litter. A similar profile in terms of research can be found in Europe where the best represented topics within European projects were 'Policy, Governance and Management' and 'Monitoring'. Comparatively 'Risk Assessment', 'Fragmentation' and 'Assessment Tools' were underrepresented (Maes et al., 2019). Several global scientific initiatives such as capacity building, technology transfer and collaborations can contribute to promote marine plastic pollution research in Africa. The African continent needs to put research effort into understanding the impacts of marine plastics specifically on human health, the economy of the continent as well as the social impacts associated with it. In order to develop policies and management strategies to aid with how to handle plastics from a manufacturing, use and reuse perspective, we need to understand the impacts holistically. However, in a continent stricken by poverty, environmental research is seldom prioritised. Public expenditure tends to focus on areas such as education, agriculture and health. For example, in the current context of the global COVID-19 pandemic, improving sanitation, hygiene and access to potable water is a high priority to reduce the spread of the virus (Jiwani & Antiporta, 2020; Marcos-Garcia et al., 2021). Improving sanitation, sewage systems and hygiene will also reduce marine litter inputs, and cross benefits should be sought where possible. Nevertheless, taking into account the current and forthcoming impacts of marine plastic litter, there is a need to address the problem with innovative measures. There is a need for knowledge transfer and capacity building to reduce plastic where possible, while implementing better waste management systems and infrastructure throughout Africa.

Acknowledgements We would like to acknowledge the valuable insights proved by Peter Ryan, Anham Salyani, Salieu Sankoh and Tony Ribbink in peer-reviewing this chapter. We would also like to acknowledge Nieves López and Federico Labanti (Studio Atlantis) for creating the illustrations.

Annex 3.1 Total Number of Marine Litter Impact Studies Published across Africa in Peer-Reviewed Journals as of December 2021

Impact	Country/region	Total number of studies	Citations
Social	South Africa	4	Preston-Whyte (2008), Ballance et al. (2000), Van Rensburg et al. (2020), Arabi and Nahman (2020)
	Accra, Ghana	1	Van Dyck et al. (2016)
Human health	South Africa	2	Naidoo et al. (2020), Made et al. (2020)
	Accra, Ghana	1	Van Dyck et al. (2016)
Environmental	South Africa	13	Mbedzi et al. (2019), Ryan et al. (2016) Reynolds and Ryan (2018), Weideman et al. (2020a, 2020b), Naidoo et al. (2016), Bakir et al. (2020), Nel and Froneman (2018), Iwalaye et al. (2021), Sparks (2020), Sparks and Immelman (2020), Best et al. (2001) Shaughnessy (1980), Cliff et al. (2002)
	Nigeria	3	Biginagwa et al. (2016), Akindelea et al. (2019), Adeogun et al. (2020)
	Kenya	1	Kosore et al. (2018)
	Mauritania, Canary Islands	1	Rodríguez et al. (2013)
	South Atlantic Ocean	1	Ryan et al. (1988)
Economic	South Africa	2̶	Ryan and Swanepoel (1996), Arabi and Nahman (2020)
	Mauritius	1	Chummun and Mathithibane (2020)
	Mozambique	1	Jones and Unsworth (2020)
	African Continent	2	Škare et al. (2021)
	sub-Saharan, Indian Ocean nations	1	Short et al. (2018)

References for Annex 3.1

Adeogun, A. O., Ibor, O. R., Khan, E. A., Chukwuka, A. V., Omogbemi, E. D., & Arukwe, A. (2020). Detection and occurrence of microplastics in the stomach of commercial fish species from a municipal water supply lake in southwestern Nigeria. *Environmental Science and Pollution Research, 27*, 31035–31045. https://doi.org/10.1007/s11356-020-09031-5

Akindele, E. O., Ehlers, S. M., & Koop, J. H. E. (2019). First empirical study of freshwater microplastics in West Africa using gastropods from Nigeria as bioindicators. *Limnologica, 78.* https://doi.org/10.1016/j.limno.2019.125708.

Arabi, S., & Nahman, A. (2020). Impacts of marine plastic on ecosystem services and economy: State of South African research. *South African Journal of Science, 116*, 1–7. https://doi.org/10.17159/sajs.2020/7695.

Bakir, A., van der Lingen, C. D., Preston-Whyte, F., Bali, A., Geja, Y., & Barry, J., et al. (2020). Microplastics in commercially important small pelagic fish species from South Africa. *Frontiers in Marine Science, 7*, 910. https://doi.org/10.3389/fmars.2020.574663.

Ballance, A., Ryan, P., & Turpie, J. (2000). How much is a clean beach worth? The impact of litter on beach users in the Cape Peninsula, South Africa. *South African Journal of Science, 96*, 210–213. https://doi.org/10.10520/AJA00382353_8975.

Best, P. B., Peddemors, V. M., Cockcroft, V. G., & Rice, N. (2001). Mortalities of right whales and related anthropogenic factors in South African waters, 1963–1998. *Journal of Cetacean Research and Management,* 171–176. https://doi.org/10.47536/jcrm.vi.293.

Biginagwa, F. J., Mayoma, B. S., Shashoua, Y., Syberg, K., & Khan, F. R. (2016). First evidence of microplastics in the African Great Lakes: Recovery from Lake Victoria Nile perch and Nile tilapi. *Journal of Great Lakes Research, 42*, 146–149. https://doi.org/10.1016/j.jglr.2015.10.012

Chummun, B. Z., & Mathithibane, M. (2020). Challenges and coping strategies of Covid-2019 in the tourism industry in mauritius. *African Journal of Hospitality, Tourism and Leisure, 9*, 810–822. https://doi.org/10.46222/AJHTL.19770720-53.

Cliff, G., Dudley, S. F. J., Ryan, P. G., & Singleton, N. (2002). Large sharks and plastic debris in KwaZulu-Natal, South Africa. *Marine & Freshwater Research, 53*, 575–581. https://doi.org/10.1071/MF01146.

Iwalaye, A. O., Moodley, K. G., & Robertson-Andersson, D. V. (2021). Water temperature and microplastic concentration influenced microplastic ingestion and retention rates in sea cucumber (Holothuria cinerascens Brandt, 1835). *Ocean Science Journal.* https://doi.org/10.1007/s12601-021-00013-3.

Jones, B. L., Unsworth, R. K. F. (2020). The perverse fisheries consequences of mosquito net malaria prophylaxis in East Africa. *Ambio* 49, 1257–1267. https://doi.org/10.1007/s13280-019-01280-0.

Kosore, C., Ojwang, L., Maghanga, J., Kamau, J., Kimeli, A., Omukoto, J., Ngisiag'e, N., Mwaluma, J., Ong'ada, H., Magori, C. & Ndirui, E. (2018). Occurrence and ingestion of microplastics by zooplankton in Kenya's marine environment: First documented evidence. *African Journal of Marine Science, 40*, 225–234. https://doi.org/10.2989/1814232X.2018.1492969.

Made, F., Kootbodien, T., Wilson, K., Tlotleng, N., Mathee, A., Ndaba, M., Kgalamono, S., & Naicker, N. (2020). Illness, self-rated health and access to medical care among waste pickers in landfill sites in Johannesburg, South Africa. *International Journal of Environmental Research and Public Health, 17*, 1–10. https://doi.org/10.3390/ijerph17072252.

Mbedzi, R., Dalu, T., Wasserman, R. J., Murungweni, F., & Cuthbert, R. N. (2019). Functional response quantifies microplastic uptake by a widespread African fish species. *Science of the Total Environment,* 134522. https://doi.org/10.1016/j.scitotenv.2019.134522.

Naidoo, T., Rajkaran, A., Sershen (2020). Impacts of plastic debris on biota and implications for human health: A South African perspective. *African Journal of Marine Science,.* 116. http://dx.doi.org/10.17159/sajs.2020/7693.

Naidoo, T., Smit, A.J., Glassom, D. (2016). Plastic ingestion by estuarine mullet Mugil cephalus (Mugilidae) in an urban harbour, KwaZulu-Natal, South Africa. *African Journal of Marine Science,* 38, 145–149. https://doi.org/10.2989/1814232X.2016.1159616.

Nel, H. A., & Froneman, P. W. (2018). Presence of microplastics in the tube structure of the reef-building polychaete Gunnarea gaimardi (Quatrefages 1848). *African Journal of Marine Science, 40*, 87–89. https://doi.org/10.2989/1814232X.2018.1443835.

Preston-Whyte, Robert. (2008). The Beach as a Liminal Space. https://doi.org/10.1002/978047075 2272.ch28.

Reynolds, C., & Ryan, P.G. (2018). Micro-plastic ingestion by waterbirds from contaminated wetlands in South Africa. *Marine Pollution Bulletin.* https://doi.org/10.1016/j.marpolbul.2017. 11.021.

Rodríguez, B., Bécares, J., Rodríguez, A., & Arcos, J.M. (2013). Incidence of entanglements with marine debris by northern gannets (*Morus bassanus*) in the non-breeding grounds. *Marine Pollution Bulletin* 75, 259–263. https://doi.org/10.1016/j.marpolbul.2013.07.003.

Ryan, P.G., Cole, G., Spiby, K., Nel, R., Osborne, A., & Perold, V. (2016). Impacts of plastic ingestion on post-hatchling loggerhead turtles off South Africa. *Marine Pollution Bulletin.* https://doi.org/10.1016/j.marpolbul.2016.04.005.

Ryan, P. G., Connell, A. D., & Gardner, B. D. (1988). Plastic ingestion and PCBs in seabirds: Is there a relationship? *Marine Pollution Bulletin.* https://doi.org/10.1016/0025-326X(88)90674-1.

Ryan, P. G., & Swanepoel, D. (1996). Cleaning beaches: Sweeping the rubbish under the carpet. *South African Journal of Science, 92*(6), 275–276.

Shaughnessy, P.D. (1980). Entanglement of cape fur seals with man-made objects. *Marine Pollution Bulletin.* https://ur.booksc.eu/journal/16929.

Short, R., Gurung, R., Rowcliffe, M., Hill, N., & Milner-Gulland, E. J. (2018). The use of mosquito nets in fisheries: A global perspective. *Plos One* 13, 1–14. https://doi.org/10.1371/journal.pone. 0191519.

Škare, M., Soriano, D. R., & Porada-Rochoń, M. (2021). Impact of COVID-19 on the travel and tourism industry. *Technological Forecasting and Social Change, 163.* https://doi.org/10.1016/j. techfore.2020.120469.

Sparks, C. (2020). Microplastics in mussels along the coast of cape town, South Africa. *Bulletin of Environment Contamination and Toxicology, 104*, 423–431. https://doi.org/10.1007/s00128-020-02809-w.

Sparks, C., & Immelman, S. (2020). Microplastics in offshore fish from the Agulhas Bank, South Africa. *Marine Pollution Bulletin, 156*, 111216. https://doi.org/10.1016/j.marpolbul.2020. 111216.

Van Dyck, I. P., Nunoo, F. K. E., & Lawson, E. T. (2016). An empirical assessment of marine debris, seawater quality and littering in Ghana. *Journal of Geoscience and Environment Protection, 04*, 21–36. https://doi.org/10.4236/gep.2016.45003.

Van Rensburg, M. L., Nkomo, S. L., & Dube, T. (2020). The 'plastic waste era'; social perceptions towards single-use plastic consumption and impacts on the marine environment in Durban, South Africa. *Applied Geography, 114*, 102132. https://doi.org/10.1016/j.apgeog.2019.102132.

Weideman, E. A., Munro, C., Perold, V., Omardien, A., & Ryan, P. G. (2020a). Ingestion of plastic litter by the sandy anemone Bunodactis reynaudi. *Environmental Pollution.* https://doi.org/10. 1016/j.envpol.2020.115543.

References

Abalansa, S., El Mahrad, B., Vondolia, G. K., Icely, J., & Newton, A. (2020). The marine plastic litter issue: A social-economic analysis. *Sustain, 12*, 1–27. https://doi.org/10.3390/su12208677

Abidli, S., Antunes, J. C., Ferreira, J. L., Lahbib, Y., Sobral, P., & Trigui El Menif, N. (2018). Microplastics in sediments from the littoral zone of the north Tunisian coast (Mediterranean Sea). *Estuarine, Coastal and Shelf Science, 205*, 1–9. https://doi.org/10.1016/j.ecss.2018.03.006

Abidli, S., Lahbib, Y., & Trigui El Menif, N. (2019). Microplastics in commercial molluscs from the lagoon of Bizerte (Northern Tunisia). *Marine Pollution Bulletin, 142*, 243–252. https://doi.org/10.1016/j.marpolbul.2019.03.048

Aboul-Gheit, A. K., Khalil, F. H., & Abdel-Moghny, T. (2006). Adsorption of spilled oil from seawater by waste plastic. *Oil & Gas Science and Technology, 61*(2), 259–268.

Adeyemi, G. A., Ayanda, I. O., & Dedeke, G. A. (2019). The interplay between sea turtle population and income generation in south-west Nigeria coastal environment. *Journal of Physics: Conference Series. Institute of Physics Publishing* (p. 012127). https://doi.org/10.1088/1742-6596/1299/1/012127

AFFIAN, K., ROBIN, M., MAANAN, M., DIGBEHI, B., DJAGOUA, E. V. & KOUAMÉ, F. 2009. Heavy metal and polycyclic aromatic hydrocarbons in Ebrié lagoon sediments, Côte d'Ivoire. Environ Monit Assess, 159, 531-41

Akindele, E. O., & Alimba, C. G. (2021). Plastic pollution threat in Africa: Current status and implications for aquatic ecosystem health. *Environmental Science and Pollution Research*. https://doi.org/10.1007/s11356-020-11736-6

Al-Masroori, H., Al-Oufi, H., McIlwain, J. L., & McLean, E. (2004). Catches of lost fish traps (ghost fishing) from fishing grounds near Muscat, Sultanate of Oman. *Fisheries Research, 69*, 407–414. https://doi.org/10.1016/j.fishres.2004.05.014

Alimi, O. S., Fadare, O. O., & Okoffo, E. D. (2021). Microplastics in African ecosystems: Current knowledge, abundance, associated contaminants, techniques, and research needs. *Science of the Total Environment*. https://doi.org/10.1016/j.scitotenv.2020.142422

Alimba, C. G., Faggio, C., Sivanesan, S., Ogunkanmi, A. L., & Krishnamurthi, K. (2021). Micro(nano)-plastics in the environment and risk of carcinogenesis: Insight into possible mechanisms. *Journal of Hazardous Materials, 416*, 126143. https://doi.org/10.1016/j.jhazmat.2021.126143

Alimi, O. S., Fadare, O. O., & Okoffo, E. D. (2021). Microplastics in African ecosystems: Current knowledge, abundance, associated contaminants, techniques, and research needs. *Science of the Total Environment, 755*. https://doi.org/10.1016/j.scitotenv.2020.142422

Allen, A. S., Seymour, A. C., & Rittschof, D. (2017). Chemoreception drives plastic consumption in a hard coral. *Marine Pollution Bulletin, 124*, 198–205. https://doi.org/10.1016/j.marpolbul.2017.07.030

Amelia, T. S. M., Khalik, W.M.A. W. M., ONG, M. C., Shao, Y. T., Pan, H.-J. & Bhubalan, K. 2021. Marine microplastics as vectors of major ocean pollutants and its hazards to the marine ecosystem and humans. Progress in Earth and Planetary Science, 8, 12.

Ammendolia, J., Saturno, J., Brooks, A.L., Jacobs, S., Jambeck, J.R., 2021. An emerging source of plastic pollution: Environmental presence of plastic personal protective equipment (PPE) debris related to COVID-19 in a metropolitan city. Environ. Pollut. 269, 116160. https://doi.org/10.1016/j.envpol.2020.116160

Andrady, A. L. (2017). The plastic in microplastics: A review. *Marine Pollution Bulletin, 119*, 12–22. https://doi.org/10.1016/j.marpolbul.2017.01.082

Arabi, S., & Nahman, A. (2020). Impacts of marine plastic on ecosystem services and economy: State of South African research. *South African Journal of Science, 116*, 1–7. https://doi.org/10.17159/sajs.2020/7695

Arienzo, M., Ferrara, L., & Trifuoggi, M. (2021). Research progress in transfer, accumulation and effects of microplastics in the oceans. *Journal of Marine Science and Engineering*. https://doi.org/10.3390/jmse9040433

Babayemi, J. O., Nnorom, I. C., Osibanjo, O., & Weber, R. (2019). Ensuring sustainability in plastics use in Africa: Consumption, waste generation, and projections. *Environmental Sciences Europe.* https://doi.org/10.1186/s12302-019-0254-5

Baderoon, G. (2009). The African Oceans—Tracing the sea as memory of slavery in South African literature and culture. *Research in African Literatures, 89–107.*

Bakir, A., Rowland, S. J., & Thompson, R. C. (2012). Competitive sorption of persistent organic pollutants onto microplastics in the marine environment. *Marine Pollution Bulletin, 64*(12), 2782–2789. https://doi.org/10.1016/j.marpolbul.2012.09.010

Bakir, A., van der Lingen, C. D., Preston-Whyte, F., Bali, A., Geja, Y., Barry, J., et al. (2020). Microplastics in commercially important small pelagic fish species from South Africa. *Frontiers in Marine Science, 7*, 910. https://doi.org/10.3389/fmars.2020.574663

Balderson, S. D., & Martin, L. E. C. (2015). *Environmental impacts and causation of "beached" Drifting Fish Aggregating Devices around Seychelles Islands: a preliminary report on data collected by Island Conservation Society.* Olhão, Port. IOTC WPEB 15. https://www.iotc.org/sites/default/files/documents/2019/08/IOTC-2019-WPEB15-37.pdf

Ballance, A., Ryan, P., & Turpie, J. (2000). How much is a clean beach worth? The impact of litter on beach users in the Cape Peninsula, South Africa. *South African Journal of Science, 96*, 210–213. https://doi.org/10.10520/AJA00382353_8975

Bamanga, A., Amaeze, N. H., & Al-Anzi, B. (2019). Comparative Investigation of total, recoverable and bioavailable fractions of sediment metals and metalloids in the Lagos Harbour and lagoon system. *Sustainability, 11*(16), 4339.

Beaumont, N. J., Aanesen, M., Austen, M. C., Börger, T., Clark, J. R., Cole, M., et al. (2019). Global ecological, social and economic impacts of marine plastic. *Marine Pollution Bulletin, 142*, 189–195. https://doi.org/10.1016/j.marpolbul.2019.03.022

Best, P. B., Peddemors, V. M., Cockcroft, V. G., & Rice, N. (2001). Mortalities of right whales and related anthropogenic factors in South African waters, 1963–1998. *Journal of Cetacean Research and Management,* 171–176. https://doi.org/10.47536/jcrm.vi.293

Biermann, L., Clewley, D., Martinez-Vicente, V., & Topouzelis, K. (2020). Finding plastic patches in coastal waters using optical satellite data. *Scientific Reports, 10.* https://doi.org/10.1038/s41598-020-62298-z

Biney, C., Amuzu, A. T., Calamari, D., Kaba, N., Mbome, I. L., Naeve, H., Ochumba, P. B., Osibbanjo, O., Radegonde, V. & Saad, M. A. 1994. Review of heavy metals in the African aquatic environment. Ecotoxicol Environ Saf, 28, 134-59.

Booth, D. (2005). Paradoxes of material culture: The political economy of surfing. In J. Nauright, & K. S. Schimmel (Ed.), *The political economy of sport. International political economy series.* Palgrave Macmillan. https://doi.org/10.1057/9780230524057_6

Bravo, M., Astudillo, J., Lancellotti, D., Luna-Jorquera, G., Valdivia, N., & Thiel, M. (2011). Rafting on abiotic substrata: Properties of floating items and their influence on community succession. *Marine Ecology Progress Series, 439*, 1–17. https://doi.org/10.3354/meps09344

Brown, J., & Macfadyen, G. (2007). Ghost fishing in European waters: Impacts and management responses. *Marine Policy, 31*, 488–504. https://doi.org/10.1016/j.marpol.2006.10.007

Brown, M., Dissanayake, A., Galloway, T., Lowe, D., & Thompson, R. (2008). Ingested microscopic plastic translocates to the circulatory system of the mussel, *mytilus edulis* (L.). *Environmental Science and Technology, 42*, 5026–5031. https://doi.org/10.1021/es800249a

Browne, M. A., Niven, S. J., Galloway, T. S., Rowland, S. J., & Thompson, R. C. (2013). Microplastic moves pollutants and additives to worms, reducing functions linked to health and biodiversity. *Current Biology, 23*, 2388–2392. https://doi.org/10.1016/j.cub.2013.10.012

Bruce-Vanderpuije, P., Megson, D., Reiner, E. J., Bradley, L., Adu-Kumi, S., & Gardella, J. A. (2019). The state of POPs in Ghana- A review on persistent organic pollutants: Environmental and human exposure. *Environmental Pollution, 245*, 331–342. https://doi.org/10.1016/j.envpol.2018.10.107

Burt, A. J., Raguain, J., Sanchez, C., Brice, J., Fleischer-Dogley, F., Goldberg, R., et al. (2020). The costs of removing the unsanctioned import of marine plastic litter to small island states. *Science and Reports, 10*, 1–11. https://doi.org/10.1038/s41598-020-71444-6

Carpenter, E. J., Anderson, S. J., Harvey, G. R., Miklas, H. P., & Peck, B. B. (1972). Polystyrene spherules in coastal waters. *Science, 178*, 749–750. https://doi.org/10.1126/science.178.4062.749

Carpenter, E. J., & Smith, K. L. (1972). Plastics on the Sargasso Sea surface. *Science, 175*, 1240–1241. https://doi.org/10.1126/science.175.4027.1240

CBD. (2016). Marine debris: Understanding, preventing and mitigating the significant adverse impacts on marine and coastal biodiversity, CBD Technical Series. https://www.cbd.int/doc/publications/cbd-ts-83-en.pdf

Chen, C. E., Liu, Y. S., Dunn, R., Zhao, J. L., Jones, K. C., Zhang, H., ... & Sweetman, A. J. (2020). A year-long passive sampling of phenolic endocrine disrupting chemicals in the East River, South China. *Environment International, 143*, 105936.

Chen, H. J., Ngowi, E. E., Qian, L., Li, T., Qin, Y. Z., Zhou, J. J., ... & Wu, D. D. (2021). Role of hydrogen sulfide in the endocrine system. *Frontiers in Endocrinology, 12*.

Chummun, B. Z., & Mathithibane, M. (2020). Challenges and coping strategies of Covid-2019 in the tourism industry in mauritius. *African Journal of Hospitality, Tourism and Leisure, 9*, 810–822. https://doi.org/10.46222/AJHTL.19770720-53

Cliff, G., Dudley, S. F. J., Ryan, P. G., & Singleton, N. (2002). Large sharks and plastic debris in KwaZulu-Natal, South Africa. *Marine & Freshwater Research, 53*, 575–581. https://doi.org/10.1071/MF01146

Common, M., & Stagl, S. (2005). *Ecological economics an introduction*. Press. https://doi.org/10.1017/CBO9780511805547

Cundell, A. M. (1974). Plastics in the marine environment. *Environmental Conservation, 1*, 63–68. https://doi.org/10.1002/etc.2426

Dasgupta, P. (2021). *The economics of biodiversity: The Dasgupta review*. Hm Treasury.

de Graaf, G., & Garibaldi, L. (2014). The value of African fisheries, FAO Fisheries and *Aquaculture Circular, 1093*. https://doi.org/10.1578/AM.40.3.2014.297

de Vos, A., Aluwihare, L., Youngs, S., DiBenedetto, M. H., Ward, C. P., Michel, A. P., ... & James, B. D. (2021). The m/v x-press pearl nurdle spill: Contamination of burnt plastic and unburnt nurdles along Sri Lanka's Beaches. *ACS Environmental Au, 2*(2), 128–135.

Derraik, J. G. B. (2002). The pollution of the marine environment by plastic debris: A review. *Marine Pollution Bulletin, 44*, 842–852. https://doi.org/10.1016/S0025-326X(02)00220-5

Diop, S., & Asongu, S. (2021). Research productivity: Trend and comparative analyses by regions and continents. *European Xtramile Centre of African Studies*. https://doi.org/10.2139/ssrn.3855361

Duhec, A. V., Jeanne, R. F., Maximenko, N., & Hafner, J. (2015). Composition and potential origin of marine debris stranded in the Western Indian Ocean on remote Alphonse Island, Seychelles. *Marine Pollution Bulletin, 96*, 76–86.

Duncan, E. M., Botterell, Z. L. R., Broderick, A. C., Galloway, T. S., Lindeque, P. K., Nuno, A., & Godley, B. J. (2017). A global review of marine turtle entanglement in anthropogenic debris: A baseline for further action. *Endangered Species Research, 34*, 431–448. https://doi.org/10.3354/esr00865

Dunlop, S. W., Dunlop, B. J., & Brown, M. (2020). Plastic pollution in paradise: Daily accumulation rates of marine litter on Cousine Island, Seychelles. *Marine Pollution Bulletin, 151*, 110803. https://doi.org/10.1016/j.marpolbul.2019.110803

Ecosystems and Human Well-being. (2005). Millenium Ecosystem Assessment, Island Pre. ed, *Ecosystems and human well-being: Synthesis*. Washington DC 20002. https://doi.org/10.5822/978-1-61091-484-0_1

Eich, A., Mildenberger, T., Laforsch, C., & Weber, M. (2015). Biofilm and diatom succession on polyethylene (PE) and biodegradable plastic bags in two marine habitats: Early signs of degradation in the pelagic and benthic zone? *PLoS ONE*. https://doi.org/10.1371/journal.pone.0137201

Ellen MacArthur Foundation. (2020). The Global Commitment 2020. https://ellenmacarthurfoun dation.org/news/global-commitment-2020-progress-report-published

Ellen MacArthur Foundation. (2017). The new plastics economy: Rethinking the future of plastics & catalysing action. *Ellen MacArthur Found, 68.* https://doi.org/10.1103/Physrevb.74.035409.

Fadare, O. O., & Okoffo, E. D. (2020). Covid-19 face masks: A potential source of microplastic fibers in the environment. *The Science of the Total Environment, 737.* https://doi.org/10.1016/j. scitotenv.2020.140279

Fazey, F. M. C., & Ryan, P. G. (2016). Biofouling on buoyant marine plastics: An experimental study into the effect of size on surface longevity. *Environmental Pollution.* https://doi.org/10. 1016/j.envpol.2016.01.026

Fernández, B., Santos-Echeandía, J., Rivera-Hernández, J.R., Garrido, S., Albentosa, M. (2020). Mercury interactions with algal and plastic microparticles: Comparative role as vectors of metals for the mussel, *Mytilus galloprovincialis. Journal of Hazardous Materials, 396,* 122739. https:// doi.org/10.1016/j.jhazmat.2020.122739

Fred-Ahmadu, O. H., Bhagwat, G., Oluyoye, I., Benson, N. U., Ayejuyo, O. O., & Palanisami, T. (2020). Interaction of chemical contaminants with microplastics: Principles and perspectives. *Science of the Total Environment, 706,* 135978.

Gall, S. C., & Thompson, R. C. (2015). The impact of debris on marine life. *Marine Pollution Bulletin, 92,* 170–179. https://doi.org/10.1016/j.marpolbul.2014.12.041

Galloway, T., Cole, M., & Lewis, C. (2017). Interactions of microplastic debris throughout the marine ecosystem. *Nature Ecology & Evolution, 1.* https://doi.org/10.1038/s41559-017-0116

GESAMP. (2015). *Report of the 42nd session of the joint group of experts on the scientific aspects of marine environmental protection (GESAMP),* IOC-UNESCO.

Gilardi, K. V. K., Antonelis, K., Galgani, F., Grilly, E., He, P., Linden, O., Piermarini, R., Richardson, K., Santillo, D., Thomas, S. N., Van den Dries, P., Wang, L. (2020). Sea-based sources of marine litter – a review of current knowledge and assessment of data gaps.

Gilman, E., Chopin, F., Suuronen, P., Kuemlangan, B. (2016). *Abandoned, lost and discarded gillnets and trammel nets. Methods to estimate ghost fishing mortality, and the status of regional monitoring and management,* FAO Fisheries Technical Paper. https://agris.fao.org/agris-search/ search.do?recordID=XF2017001196

Godswill, C., Gospel, C. (2019). Impacts of plastic pollution on the sustainability of seafood value chain and human health. *International Journal of Advance Scientific Research, 5,* 46–138. https://www.ijaar.org/articles/Volume5-Number11/Sciences-Technology-Engineering/ ijaar-ste-v5n11-nov19-p1.pdf

Gregory, M. R. (2009). Environmental implications of plastic debris in marine settings-entanglement, ingestion, smothering, hangers-on, hitch-hiking and alien invasions. *Philosophical Transactions of the Royal Society B: Biological Sciences.* https://doi.org/10.1098/rstb.2008.0265.

Guzzetti, E., Sureda, A., Tejada, S., & Faggio, C. (2018). Microplastic in marine organism: Environmental and toxicological effects. *Environmental Toxicology and Pharmacology, 64,* 164–171. https://doi.org/10.1016/j.etap.2018.10.009

Haines-Young, R., & Potschin, M. (2012). *Common international classification of ecosystem services.* Centre for Environmental Management, University of Nottingham, Nottingham.

Hall, N. M., Berry, K. L. E., Rintoul, L., & Hoogenboom, M. O. (2015). Microplastic ingestion by scleractinian corals. *Marine Biology, 162,* 725–732. https://doi.org/10.1007/s00227-015-2619-7

Hamilton, L.A., Feit, S., Muffett, C., & Kelso, M. (2019). Plastic & Climate: The hidden costs of plastic planet. *Center for International Environmental Law (CIEL),* 1–108. https://greenwire.gre enpeace.de/system/files/2019-05/20190515-report-plastic-and-climate-ciel.pdf

Harris, L. S. T., Gill, H., & Carrington, E. (2021). Microplastic changes the sinking and resuspension rates of marine mussel biodeposits. *Marine Pollution Bulletin, 165,* 112165. https://doi.org/10. 1016/j.marpolbul.2021.112165

Heyerdahl, T. (1971). Atlantic ocean pollution and biota observed by the "Ra" expeditions. *Biological Conservation, 3,* 164–167. https://doi.org/10.1016/0006-3207(71)90158-3

Hierl, F., Wu, H. C., & Westphal, H. (2021). Scleractinian corals incorporate microplastic particles: Identification from a laboratory study. *Environmental Science and Pollution Research.* https://doi.org/10.1007/s11356-021-13240-x

Hosoda, J., Ofosu-Anim, J., Sabi, E. B., Akita, L. G., Onwona-Agyeman, S., Yamashita, R., & Takada, H. (2014). Monitoring of organic micropollutants in Ghana by combination of pellet watch with sediment analysis: E-waste as a source of PCBs. *Marine Pollution Bulletin, 86,* 575–581. https://doi.org/10.1016/j.marpolbul.2014.06.008

Indian Ocean Comission. (2021a). Gestion des Déchets.

Indian Ocean Comission. (2021b). ExPLOI.

Indian Ocean Comission. (2021c). SWIOFISH2.

International Monetary Fund. (2021). South Africa GDP (Current Prices). https://www.imf.org/en/Countries/ZAF

ITOPF. (2020). The International Tanker Owners Pollution Federation Limited Oil Tanker Spill Statistics.

Iwalaye, A. O., Moodley, K. G., & Robertson-Andersson, D. V. (2021). Water temperature and microplastic concentration influenced microplastic ingestion and retention rates in sea cucumber (Holothuria cinerascens Brandt, 1835). *Ocean Science Journal.* https://doi.org/10.1007/s12601-021-00013-3

Iwalaye, O. A., Moodley, G. K., & Robertson-Andersson, D. V. (2020). The possible routes of microplastics uptake in sea cucumber Holothuria cinerascens (Brandt, 1835). *Environmental Pollution, 264,* 114644. https://doi.org/10.1016/j.envpol.2020.114644

Jiwani, S. S., & Antiporta, D. A. (2020). Inequalities in access to water and soap matter for the COVID-19 response in sub-Saharan Africa. *International Journal for Equity in Health, 19*(1), 1–3.

Jones, B. L., & Unsworth, R. K. F. (2020). The perverse fisheries consequences of mosquito net malaria prophylaxis in East Africa. *Ambio, 49,* 1257–1267. https://doi.org/10.1007/s13280-019-01280-0

Kadafa, A. A. (2012). Environmental impacts of oil exploration and exploitation in the Niger Delta of Nigeria. *Global Journal of Science Frontier Research Environment & Earth Sciences, 12*(3), 19–28.

Karamanlidis, A. A., Androukaki, E., Adamantopoulou, S., Chatzispyrou, A., Johnson, W. M., Kotomatas, S., Papadopoulos, A., Paravas, V., Paximadis, G., Pires, R. & Tounta, E. (2008). Assessing accidental entanglement as a threat to the Mediterranean monk seal *Monachus monachus. Endangered Species Research, 5,* 205–213. https://doi.org/10.3354/esr00092

Karlsson, T. M., Vethaak, A. D., Almroth, B. C., Ariese, F., van Velzen, M., Hassellöv, M., & Leslie, H. A. (2017). Screening for microplastics in sediment, water, marine invertebrates and fish: Method development and microplastic accumulation. *Marine Pollution Bulletin, 122,* 403–408. https://doi.org/10.1016/j.marpolbul.2017.06.081

Keswani, A., Oliver, D. M., Gutierrez, T., & Quilliam, R. S. (2016). Microbial hitchhikers on marine plastic debris: Human exposure risks at bathing waters and beach environments. *Marine Environment Research, 118,* 10–19. https://doi.org/10.1016/j.marenvres.2016.04.006

Kiessling, T., Gutow, L., & Thiel, M. (2015). Marine litter as habitat and dispersal vector BT. *Marine Anthropogenic Litter* 141–181. https://doi.org/10.1007/978-3-319-16510-3_6.

Koelmans, A. A., Bakir, A., Burton, G. A., & Janssen, C. R. (2016). Microplastic as a vector for chemicals in the aquatic environment: Critical review and model-supported reinterpretation of empirical studies. *Environmental Science & Technology, 50*(7), 3315–3326. https://doi.org/10.1021/acs.est.5b06069

Kosore, C., Ojwang, L., Maghanga, J., Kamau, J., Kimeli, A., Omukoto, J., Ngisiag'e, N., Mwaluma, J., Ong'ada, H., Magori, C. & Ndirui, E. (2018). Occurrence and ingestion of microplastics by zooplankton in Kenya's marine environment: First documented evidence. *African Journal of Marine Science, 40,* 225–234. https://doi.org/10.2989/1814232X.2018.1492969

Krystosik, A., Njoroge, G., Odhiambo, L., Forsyth, J. E., Mutuku, F., & LaBeaud, A. D. (2020). Solid wastes provide breeding sites, burrows, and food for biological disease vectors, and urban zoonotic reservoirs: A call to action for solutions-based research. *Frontiers in Public Health, 7*, 1–17. https://doi.org/10.3389/fpubh.2019.00405

Kühn, S., Bravo Rebolledo, E. L., & van Franeker, J. A. (2015). Deleterious effects of litter on marine life BT. *Marine Anthropogenic Litter,* 75–116. https://doi.org/10.1007/978-3-319-16510-3_4

Lachmann, F., Almroth, B.C., Baumann, H., Broström, G., Corvellec, H., Gipperth, L. (2017). Marine plastic litter on small island developing states (Sids): Impacts and measures. *The Swedish Institute for the Marine Environment, 4*, 1–76. https://havsmiljoinstitutet.se/digitalAssets/1641/1641020_sime-2017-4-marine-plastic-litter.pdf

Laist, D. (1987). Bio effects of lost and discarded plastics on marine Biota (Good Opening Line). *Marine Pollution Bulletin, 18*, 319–326. https://doi.org/10.1016/S0025-326X(87)80019-X

Laist, D. W. (1997). Impacts of marine debris: Entanglement of marine life in marine debris including a comprehensive list of species with entanglement and ingestion records. *Marine Debris.* https://doi.org/10.1007/978-1-4613-8486-1_10

Landrigan, P. J., Stegeman, J. J., Fleming, L. E., Allemand, D., Anderson, D. M., Backer, L. C., Brucker-Davis, F., Chevalier, N., Corra, L., Czerucka, D., & Bottein, M. Y. D. (2020). Human health and ocean pollution. *Annals of Global Health, 86*, 1–64.

Lavers, J. L., Rivers-Auty, J., & Bond, A. L. (2021). Plastic debris increases circadian temperature extremes in beach sediments. *Journal of Hazardous Materials, 416*. https://doi.org/10.1016/j.jhazmat.2021.126140

Lebreton, L., Slat, B., Ferrari, F., Sainte-Rose, B., Aitken, J., Marthouse, R., ... & Reisser, J. (2018). Evidence that the Great Pacific Garbage Patch is rapidly accumulating plastic. *Scientific Reports, 8*(1), 1–15.

Lewis, D. (2020). How Mauritius is cleaning up after major oil spill in biodiversity hotspot. *Nature, 585*(7824), 172–173.

Lobelle, D., & Cunliffe, M. (2011). Early microbial biofilm formation on marine plastic debris. *Marine Pollution Bulletin, 62*, 197–200. https://doi.org/10.1016/j.marpolbul.2010.10.013

Lusher, A. (2015). Microplastics in the marine environment: Distribution, interactions and effects BT. *Marine Anthropogenic Litter,* 245–307. https://doi.org/10.1007/978-3-319-16510-3_10

Lusher, A. L., Welden, N. A., Sobral, P., & Cole, M. (2017). Sampling, isolating and identifying microplastics ingested by fish and invertebrates. *Analytical Methods, 9*, 1346–1360. https://doi.org/10.1039/c6ay02415g

Maanan, M. (2008). Heavy metal concentrations in marine molluscs from the Moroccan coastal region. *Environmental Pollution, 153*(1), 176–183.

Macfadyen, G., Huntington, T., & Cappell, R. (2009). *Abandoned, lost or otherwise discarded fishing gear*. UNEP Regional Seas Reports and Studies No.185; FAO fisheries and aquaculture technical paper, No. 523. Rome, UNEP/FAO. 2009. 115p.

Made, F., Kootbodien, T., Wilson, K., Tlotleng, N., Mathee, A., Ndaba, M., Kgalamono, S., & Naicker, N. (2020). Illness, self-rated health and access to medical care among waste pickers in landfill sites in Johannesburg, South Africa. *International Journal of Environmental Research and Public Health, 17*, 1–10. https://doi.org/10.3390/ijerph17072252

Maes, T., Barry, J., Stenton, C., Roberts, E., Hicks, R., Bignell, J. (2020a). The world is your oyster: low-dose, long-term microplastic exposure of juvenile oysters. *Heliyon* 6. https://doi.org/10.1016/j.heliyon.2019.e03103

Maes, T., van Diemen de Jel, J., Vethaak, A. D., Desender, M., Bendall, V. A., Van Velzen, M., & Leslie, H. A. (2020b). You are what you eat, microplastics in porbeagle sharks from the north east atlantic: Method development and analysis in spiral valve content and tissue. *Frontiers in Marine Science, 0*, 273. https://doi.org/10.3389/FMARS.2020b.00273

Maes, T., Perry, J., Alliji, K., Clarke, C., & Birchenough, S. N. (2019). Shades of grey: marine litter research developments in Europe. *Marine Pollution Bulletin, 146*, 274–281.

Mansour, S. A. (2009). Persistent organic pollutants (POPs) in Africa: Egyptian scenario. *Human & Experimental Toxicology*, 531–566. https://doi.org/10.1177/096032710934704

Marcos-Garcia, P., Carmona-Moreno, C., López-Puga, J., & García, A. R. R. (2021). COVID-19 pandemic in Africa: Is it time for water, sanitation and hygiene to climb up the ladder of global priorities?. *Science of the Total Environment, 791*, 148252.

Mbedzi, R., Dalu, T., Wasserman, R. J., Murungweni, F., & Cuthbert, R. N. (2019). Functional response quantifies microplastic uptake by a widespread African fish species. *Science of the Total Environment,* 134522. https://doi.org/10.1016/j.scitotenv.2019.134522.

Mboera, L. E., Mweya, C. N., Rumisha, S. F., Tungu, P. K., Stanley, G., Makange, M. R., Misinzo, G., De Nardo, P., Vairo, F., & Oriyo, N. M. (2016). The risk of dengue virus Tanzania. *PLoS Neglected Tropical Diseases, 10*, 15. https://doi.org/10.1371/journal.pntd.0004313

McGregor, S., & Strydom, N. A. (2020). Feeding ecology and microplastic ingestion in *Chelon richardsonii* (Mugilidae) associated with surf diatom *Anaulus australis* accumulations in a warm temperate South African surf zone. *Marine Pollution Bulletin, 158*, 111430. https://doi.org/10.1016/j.marpolbul.2020.111430

Mearns, A. J., Reish, D. J., Oshida, P. S., Ginn, T., Rempel-Hester, M. A., Arthur, C., Rutherford, N., & Pryor, R. (2018). Effects of pollution on marine organisms. *Water Environment Research, 90*. https://doi.org/10.2175/106143018X15289915807218

Menéndez-Pedriza, A., & Jaumot, L. (2020). Microplastics: A critical review of sorption factors. *Toxics, 8*, 1–40.

Moore, C. J. (2008). Synthetic polymers in the marine environment: A rapidly increasing, long-term threat. *Environmental Research, 108*, 131–139. https://doi.org/10.1016/j.envres.2008.07.025

Moore, S. L., Gregorio, D., Carreon, M., Weisberg, S. B., & Leecaster, M. K. (2001). Composition and distribution of beach Debris in Orange County California. *Marine Pollution Bulletin, 42*, 241–245. https://doi.org/10.1016/S0025-326X(00)00148-X

Moore, R. E., Millar, B. C., & Moore, J. E. (2020). Antimicrobial resistance (AMR) and marine plastics: Can food packaging litter act as a dispersal mechanism for AMR in oceanic environments?. *Marine Pollution Bulletin, 150*, 110702.

Mouat, J., Lozano, R. L., & Bateson, H. (2010). Economic impacts of marine litter. *Kommunenes Internasjonale Miljøorganisasjon.* https://www.kimointernational.org/wp/wp-content/uploads/2017/09/KIMO_Economic-Impacts-of-Marine-Litter.pdf

Naidoo, T., Rajkaran, & A., Sershen (2020). Impacts of plastic debris on biota and implications for human health: A South African perspective. *South African Journal of Science, 116*. https://doi.org/10.17159/sajs.2020/7693

Naidoo, T., Smit, A. J., & Glassom, D. (2016). Plastic ingestion by estuarine mullet *Mugil cephalus* (Mugilidae) in an urban harbour, KwaZulu-Natal, South Africa. *African Journal of Marine Science, 38*, 145–149. https://doi.org/10.2989/1814232X.2016.1159616

Naik, R. K., Naik, M. M., D'Costa, P. M., & Shaikh, F. (2019). Microplastics in ballast water as an emerging source and vector for harmful chemicals, antibiotics, metals, bacterial pathogens and HAB species: A potential risk to the marine environment and human health. *Marine Pollution Bulletin.* https://doi.org/10.1016/j.marpolbul.2019.110525

Näkki, P., Eronen-Rasimus, E., Kaartokallio, H., Kankaanpää, H., Setälä, O., Vahtera, E., & Lehtiniemi, M. (2021). Polycyclic aromatic hydrocarbon sorption and bacterial community composition of biodegradable and conventional plastics incubated in coastal sediments. *Science of the Total Environment, 755*, 143088.

Napper, I. E., & Thompson, R. C. (2020). Plastic debris in the marine environment: history and future challenges. *Global Challenges, 4*, 1900081. https://doi.org/10.1002/gch2.201900081

Nel, H. A., & Froneman, P. W. (2018). Presence of microplastics in the tube structure of the reef-building polychaete *Gunnarea gaimardi* (Quatrefages 1848). *African Journal of Marine Science, 40*, 87–89. https://doi.org/10.2989/1814232X.2018.1443835

Newman, S., Watkins, E., Farmer, A., Brink, P. Ten, & Schweitzer, J. P. (2015). The economics of marine litter. In *Marine anthropogenic litter* (pp. 367–394). https://doi.org/10.1007/978-3-319-16510-3_14

Nicholls, R. J., & Small, C. (2002). Improved estimates of coastal population and exposure to hazards released. *Eos, Transactions American Geophysical Union, 83*(28), 301–305.

Okuku, E., Kiteresi, L., Owato, G., Otieno, K., Mwalugha, C., Mbuche, M., ... & Omire, J. (2021). The impacts of COVID-19 pandemic on marine litter pollution along the Kenyan Coast: A synthesis after 100 days following the first reported case in Kenya. *Marine Pollution Bulletin, 162*. https://doi.org/10.1016/j.marpolbul.2020.111840

Onink, V., Jongedijk, C. E., Hoffman, M. J., van Sebille, E., & Laufkötter, C. (2021). Global simulations of marine plastic transport show plastic trapping in coastal zones. *Environmental Research Letters, 16*, 064053. https://doi.org/10.1088/1748-9326/abecbd

Oosterhuis, F., Papyrakis, E., & Boteler, B. (2015). Ocean and coastal management economic instruments and marine litter control. *Ocean and Coastal Management, 102*, 47–54. https://doi.org/10.1016/j.ocecoaman.2014.08.005

Orr, K. K., Burgess, J. E., & Froneman, P. W. (2008). The effects of mouth-phase and rainfall on the concentration of heavy metals in the sediment and water of select Eastern Cape estuaries, South Africa. *Water SA, 34*, 39–52.

Piccardo, M., Renzi, M., & Terlizzi, A. (2020). Nanoplastics in the oceans: Theory, experimental evidence and real world. *Marine Pollution Bulletin, 157*, 111317. https://doi.org/10.1016/j.marpolbul.2020.111317

Plastic Europe. (2019). *Plastics—The facts 2019*. https://plasticseurope.org/wp-content/uploads/2021/10/2019-Plastics-the-facts.pdf

Preston-Whyte, R. (2008). *The beach as a liminal space*. https://doi.org/10.1002/9780470752272.ch28

Provencher, J. F., Bond, A. L., Avery-Gomm, S., Borrelle, S. B., Rebolledo, E. L. B., Hammer, S., Kühn, S., Lavers, J. L., Mallory, M. L., Trevail, A., & Van Franeker, J. A. (2017). Quantifying ingested debris in marine megafauna: A review and recommendations for standardization. *Analytical Methods, 9*, 1454–1469. https://doi.org/10.1039/c6ay02419j

Randall, P. (2020). South African marine fisheries and abandoned, lost and discarded fishing gear. *Commonwealth Litter Programme-South Africa*. https://doi.org/10.13140/RG.2.2.30135.96162

Rasool, F. N., Saavedra, M. A., Pamba, S., Perold, V., Mmochi, A. J., Maalim, M., Simonsen, L., Buur, L., Pedersen, R. H., Syberg, K. & Jelsbak, L. (2021). Isolation and characterization of human pathogenic multidrug resistant bacteria associated with plastic litter collected in Zanzibar. *Journal of Hazardous Materials, 405*, 124591. https://doi.org/10.1016/J.JHAZMAT.2020.124591

Reusch, K., Suárez, N., Ryan, P. G., & Pichegru, L. (2020). Foraging movements of breeding Kelp Gulls in South Africa. *Movement Ecology, 81*(8), 1–12. https://doi.org/10.1186/S40462-020-00221-X

Reynolds, C., & Ryan, P. G. (2018). Micro-plastic ingestion by waterbirds from contaminated wetlands in South Africa. *Marine Pollution Bulletin*. https://doi.org/10.1016/j.marpolbul.2017.11.021

Richardson, K., Hardesty, B. D., & Wilcox, C. (2019). Estimates of fishing gear loss rates at a global scale: A literature review and meta-analysis. *Fish and Fisheries, 20*, 1218–1231. https://doi.org/10.1111/faf.12407

Rochman, C. M. (2015). The complex mixture, fate and toxicity of chemicals associated with plastic debris in the marine environment. *Marine Anthopogenic Literaturer,* 117–140. https://doi.org/10.1007/978-3-319-16510-3_5

Rochman, C. M., Kurobe, T., Flores, I., & Teh, S. J. (2014). Early warning signs of endocrine disruption in adult fish from the ingestion of polyethylene with and without sorbed chemical pollutants from the marine environment. *Science of the Total Environment, 493*, 656–661.

Rochman, C. M., Brookson, C., Bikker, J., Djuric, N., Earn, A., Bucci, K., ... & Hung, C. (2019). Rethinking microplastics as a diverse contaminant suite. *Environmental Toxicology and Chemistry, 38*(4), 703–711.

Rodríguez, B., Bécares, J., Rodríguez, A., & Arcos, J. M. (2013). Incidence of entanglements with marine debris by northern gannets (*Morus bassanus*) in the non-breeding grounds. *Marine Pollution Bulletin, 75*, 259–263. https://doi.org/10.1016/j.marpolbul.2013.07.003

Rossi, G., Barnoud, J., & Monticelli, L. (2014). Polystyrene nanoparticles perturb lipid membranes. *Journal of Physical Chemistry Letters, 5*, 46. https://doi.org/10.1021/jz402234c

Ryan, P. G. (2008). Seabirds indicate changes in the composition of plastic litter in the Atlantic and south-western Indian Oceans. *Marine Pollution Bulletin, 56*, 1406–1409. https://doi.org/10.1016/j.marpolbul.2008.05.004

Ryan, P. G. (2015). A brief history of marine litter research. *Marine Anthropogenic Litter,* 1–25. https://doi.org/10.1007/978-3-319-16510-3_1

Ryan, P. G. (2016). Ingestion of plastics by marine organisms. In: Takada, H., Karapanagioti, H. K. (Eds.), *Hazardous chemicals associated with plastics in the environment. Handbook of Environmental Chemistry, 78*, 235–266. https://doi.org/10.1007/698_2016_21

Ryan, P. G. (2018). Entanglement of birds in plastics and other synthetic materials. *Marine Pollution Bulletin.* https://doi.org/10.1016/j.marpolbul.2018.06.057

Ryan, P. G. (2020a). The transport and fate of marine plastics in South Africa and adjacent oceans. *South African Journal of Science, 116*, 9. https://doi.org/10.17159/sajs.2020a/7677

Ryan, P. G. (2020b). Land or sea? What bottles tell us about the origins of beach litter in Kenya. *Waste Management, 116*, 49–57. https://doi.org/10.1016/j.wasman.2020.07.044

Ryan, P. G. (1990). The marine plastic debris problem off southern Africa: Types of debris, their environmental effects, and control measures. In *Proceedings of the Second International Conference on Marine Debris. NOAA Tech. Memo.* NMFS-SWFSC-154.

Ryan, P. G. (1988). The characteristics and distribution of plastic particles at the sea-surface off the southwestern Cape Province, South Africa. *Marine Environmental Research.* https://doi.org/10.1016/0141-1136(88)90015-3

Ryan, P. G., Bouwman, H., Moloney, C. L., Yuyama, M., & Takada, H. (2012). Long-term decreases in persistent organic pollutants in South African coastal waters detected from beached polyethylene pellets. *Marine Pollution Bulletin.* https://doi.org/10.1016/j.marpolbul.2012.09.013

Ryan, P. G., Cole, G., Spiby, K., Nel, R., Osborne, A., & Perold, V. (2016). Impacts of plastic ingestion on post-hatchling loggerhead turtles off South Africa. *Marine Pollution Bulletin.* https://doi.org/10.1016/j.marpolbul.2016.04.005

Ryan, P. G., Maclean, K., Weideman, E. A. (2020a). The Impact of the COVID-19 Lockdown on Urban Street Litter in South Africa. *Environmental Processes, 7*, 1302–1312. https://doi.org/0.1007/s40710-020-00472-1

Ryan, P. G., Pichegru, L., Perold, V., & Moloney, C. L. (2020b). Monitoring marine plastics-will we know if we are making a difference? *South African Journal of Science, 116.* https://doi.org/10.17159/sajs.2020b/7678

Ryan, P. G., & Swanepoel, D. (1996). Cleaning beaches: Sweeping the rubbish under the carpet. *South African Journal of Science, 92*(6), 275–276.

Sambyal, S. S. (2018). Five African countries among top 20 highest contributors to plastic marine debris in the world. *Down to Earth.* https://www.downtoearth.org.in/news/waste/when-oceans-fill-apart-60629

Sancho, G., Puente, E., Bilbao, A., Gomez, E., & Arregi, L. (2003). Catch rates of monkfish (*Lophius spp.*) by lost tangle nets in the Cantabrian Sea (northern Spain). *Fisheries Research, 64*, 129–139. https://doi.org/10.1016/S0165-7836(03)00212-1

Santos, M. N., Saldanha, H., Gaspar, M. B., & Monteiro, C. C. (2003). Causes and rates of net loss off the Algarve (Southern Portugal). *Fisheries Research, 64*, 115–118. https://doi.org/10.1016/S0165-7836(03)00210-8

Schleyer, M. H., & Tomalin, B. J. (2000). Damage on South African coral reefs and an assessment of their sustainable diving capacity using a fisheries approach. *Bulletin of Marine Science, 67*, 1025–1042.

Scutariu, R.E., Puiu, D., Nechifor, G., Niculescu, M., Pascu, L.F., & Galaon, T. (2019). In vitro sorption study of some organochlorine pesticides on polyethylene terephthalate microplastics. *Revista de Chimie, 70*, 4620–4626. https://doi.org/10.37358/RC.19.12.7803

Selvaranjan, K., Navaratnam, S., Rajeev, P., & Ravintherakumaran, N. (2021). Environmental challenges induced by extensive use of face masks during COVID-19: A review and potential solutions. *Environmental Challenges, 3*, 100039.

Seuront, L., Nicastro, K. R., McQuaid, C. D., & Zardi, G. I. (2021). Microplastic leachates induce species-specific trait strengthening in intertidal mussels. *Ecological Applications, 31*. https://doi. org/10.1002/eap.2222

Sevwandi Dharmadasa, W. S., Andrady, A. L., Kumara, P. T. P., Maes, T., & Gangabadage, C. S. (2021). Microplastic pollution in marine protected areas of Southern Sri Lanka. *Marine Pollution Bulletin, 168*, 112462.

Shabaka, S. H., Ghobashy, M., & Marey, R. S. (2019). Identification of marine microplastics in Eastern Harbor, Mediterranean Coast of Egypt, using differential scanning calorimetry. *Marine Pollution Bulletin, 142*, 494–503. https://doi.org/10.1016/j.marpolbul.2019.03.062

Shaughnessy, P.D. (1980). Entanglement of cape fur seals with man-made objects. *Marine Pollution Bulletin.* https://ur.booksc.eu/journal/16929

Sheavly, S. B., & Register, K. M. (2007). Marine debris & plastics: Environmental concerns, sources, impacts and solutions. *Journal of Polymers and the Environment, 15*, 301–305. https://doi.org/ 10.1007/s10924-007-0074-3

Short, R., Gurung, R., Rowcliffe, M., Hill, N., & Milner-Gulland, E. J. (2018). The use of mosquito nets in fisheries: A global perspective. *PLoS ONE, 13*, 1–14. https://doi.org/10.1371/journal.pone. 0191519

Shukla, A., Rajalakshmi, A., Subash, K., Jayakumar, S., Arul, N., Srivastava, P. K., ... & Krishnan, J. (2020). Seasonal variations of dengue vector mosquitoes in rural settings of Thiruvarur district in Tamil Nadu, India. *Journal of Vector Borne Diseases, 57*(1), 63.

Škare, M., Soriano, D. R., & Porada-Rochoń, M. (2021). Impact of COVID-19 on the travel and tourism industry. *Technological Forecasting and Social Change, 163*. https://doi.org/10.1016/j. techfore.2020.120469

Small, C., & Nicholls, R. J. (2003). A global analysis of human settlement in coastal zones. *Journal of Coastal Research*, 584–599.

Song, X., Lyu, M., Zhang, X., Ruthensteiner, B., Ahn, I. Y., Pastorino, G., et al. (2021). Large plastic debris dumps: New biodiversity hot spots emerging on the deep-sea floor. *Environmental Science & Technology Letters, 8*, 148–154. https://doi.org/10.1021/acs.estlett.0c00967

Sparks, C. (2020). Microplastics in mussels along the coast of cape town, South Africa. *Bulletin of Environment Contamination and Toxicology, 104*, 423–431. https://doi.org/10.1007/s00128-020-02809-w

Sparks, C., & Immelman, S. (2020). Microplastics in offshore fish from the Agulhas Bank, South Africa. *Marine Pollution Bulletin, 156*, 111216. https://doi.org/10.1016/j.marpolbul.2020. 111216.

Sparks, C., Odendaal, J., & Snyman, R. (2014). An analysis of historical Mussel Watch Programme data from the west coast of the Cape Peninsula, Cape Town. *Marine Pollution Bulletin, 87*, 374–380. https://doi.org/10.1016/j.marpolbul.2014.07.047

Sparks, C., Awe, A., & Maneveld, J. (2021). Abundance and characteristics of microplastics in retail mussels from Cape Town, South Africa. *Marine Pollution Bulletin.* https://doi.org/10.1016/j.mar polbul.2021.112186

Stelfox, M., Hudgins, J., & Sweet, M. (2016). A review of ghost gear entanglement amongst marine mammals, reptiles and elasmobranchs. *Marine Pollution Bulletin.* https://doi.org/10.1016/j.mar polbul.2016.06.034

Sukhsangchan, C., Phuynoi, S., Monthum, Y., Whanpetch, N., & Kulanujaree, N. (2020). Catch composition and estimated economic impacts of ghost-fishing squid traps near Suan Son Beach, Rayong province, Thailand. *Science Asia 46*, 87. https://doi.org/10.2306/scienceasia1513-1874. 2020.014

Sussarellu, R., Suquet, M., Thomas, Y., Lambert, C., Fabioux, C., Pernet, M. E. J., et al. (2016). Oyster reproduction is affected by exposure to polystyrene microplastics. *Proceedings of the National Academy of Sciences U. S. A., 113*, 2430–2435. https://doi.org/10.1073/pnas.151901 9113

Sustainable Seas Trust. (2021). *Ghost gear.* https://sst.org.za/projects/african-marine-waste-net work/research/ghost-gear/

Tall, A., Purves, M., Josupeit, H. (2016). The Pan-African fisheries and aquaculture policy framework and reform strategy: African fisheries and aquaculture in the macro economy African fisheries and the continent's natural capital. https://nepad.org/file-download/download/public/15742

Ten Brink, P., Lutchman, I., Bassi, S., Speck, S., Sheavly, S., Register, K., & Woolaway, C. (2009). *Guidelines on the use of market-based instruments to address the problem of marine litter*. Institute for European Environmental Policy (IEEP). Virginia Beach, Virginia, USA. https://wedocs.unep.org/handle/20.500.11822/2435

Thiel, M., de Veer, D., Espinoza-Fuenzalida, N. L., Espinoza, C., Gallardo, C., Hinojosa, I. A., et al. (2021). COVID lessons from the global south—Face masks invading tourist beaches and recommendations for the outdoor seasons. *Science of the Total Environment, 786*, 147486. https://doi.org/10.1016/j.scitotenv.2021.147486

Tsagbey, S. A., Mensah, A. M., & Nunoo, F. K. E. (2009). Influence of tourist pressure on beach Litter and microbial quality—Case study of two beach resorts in Ghana. *West African Journal of Applied Ecology, 15*. https://doi.org/10.4314/wajae.v15i1.49423

Turpie, J., Letley, G., Ngoma, Y., Moore, K. (2019). *The case for banning single-use plastic products in Malawi*. https://www.efdinitiative.org/publications/case-banning-single-use-plastics-malawi

UNEP (2013). *African environment outlook 3: Our environment, our health*. UNEP.

UNEP. (2021). From pollution to solution. A global assessment of marine litter and plastic pollution. In *New Scientist*. Nairobi. https://doi.org/10.1016/S0262-4079(18)30486-X

UNEP. (2016). Marine plastic debris and microplastics—Global lessons and research to inspire action and guide policy change. https://wedocs.unep.org/handle/20.500.11822/7720

UNEP. (1999). Regional overview of land-based sources and activities affecting the marine, coastal and associated freshwater environment in the west and central African Region. https://www.ais.unwater.org/ais/aiscm/getprojectdoc.php?docid=4007

UNEP/GPA. (2006). Protecting coastal and marine environment from impacts of land-based activities: A guide for national action. Trinidad and Tobago national programme of action for the protection of the coastal and marine environment from land-based sources and activities 2008–2013.

UN Environment. (2017). *UN Environment annual report*.

Van der Meulen, M. D., Devriese, L., Lee, J., Maes, T., Van Dalfsen, J. A., Huvet, A., ... & Vethaak, A. D. (2014). Socio-economic impact of microplastics in the 2 Seas, Channel and France Manche Region. https://doi.org/10.13140/RG.2.1.4487.4082

van der Mheen, M., van Sebille, E., & Pattiaratchi, C. (2020). Beaching patterns of plastic debris along the Indian Ocean rim. *Ocean Science*, 1–31. https://doi.org/10.5194/os-2020-50

Van Dyck, I. P., Nunoo, F. K. E., & Lawson, E. T. (2016). An empirical assessment of marine debris, seawater quality and littering in Ghana. *Journal of Geoscience and Environment Protection, 04*, 21–36. https://doi.org/10.4236/gep.2016.45003

Van Rensburg, M. L., Nkomo, S. L., & Dube, T. (2020). The 'plastic waste era'; Social perceptions towards single-use plastic consumption and impacts on the marine environment in Durban, South Africa. *Applied Geography, 114*, 102132. https://doi.org/10.1016/j.apgeog.2019.102132

Van Sebille, E., Aliani, S., Law, K. L., Maximenko, N., Alsina, J. M., Bagaev, A., Bergmann, M., Chapron, B., Chubarenko, I., Cózar, A., & Delandmeter, P. (2020). The physical oceanography of the transport of floating marine debris. *Environmental Research Letters, 15*. https://doi.org/10.1088/1748-9326/ab6d7d

Viool, V., Gupta, A., Petten, L., & Schalekamp, J. (2019). The price tag of plastic pollution—An economic assessment of river plastic. *Deloitte* 1–16. https://www2.deloitte.com/content/dam/Deloitte/nl/Documents/strategy-analytics-and-ma/deloitte-nl-strategy-analytics-and-ma-the-price-tag-of-plastic-pollution.pdf

Wakkaf, T., El Zrelli, R., Kedzierski, M., Balti, R., Shaiek, M., Mansour, L., Tlig-Zouari, S., Bruzaud, S., & Rabaoui, L. (2020). Microplastics in edible mussels from a southern Mediterranean lagoon: Preliminary results on seawater-mussel transfer and implications for environmental protection and seafood safety. *Marine Pollution Bulletin*. https://doi.org/10.1016/j.marpolbul.2020.111355

Weideman, E. A., Munro, C., Perold, V., Omardien, A., & Ryan, P. G. (2020a). Ingestion of plastic litter by the sandy anemone *Bunodactis reynaudi*. *Environmental Pollution*. https://doi.org/10.1016/j.envpol.2020.115543

Weideman, E. A., Perold, V., Omardien, A., Smyth, L. K., & Ryan, P. G. (2020b). Quantifying temporal trends in anthropogenic litter in a rocky intertidal habitat. *Marine Pollution Bulletin*. https://doi.org/10.1016/j.marpolbul.2020.111543

Werner, S., Budziak, A., Van Franeker, J. A., Galgani, F., Hanke, G., Maes, T., Matiddi, M., Nilsson, P., Oosterbaan, L., Priestland, E., & Thompson, R. (2016). *Harm Caused by Marine Litter*. https://doi.org/10.2788/19937

WHO, 2019. *Microplastics in drinking-water*. Geneva. https://www.who.int/news/item/20-08-2019-microplastics-in-drinking-water

Willis, K., Maureaud, C., Wilcox, C., & Hardesty, B. D. (2018). How successful are waste abatement campaigns and government policies at reducing plastic waste into the marine environment? *Marine Policy, 96*, 243–249. https://doi.org/10.1016/j.marpol.2017.11.037

World Bank Group. (2013). *SEYCHELLES tourism sector review: Sustaining growth in a successful tourism destination*. https://openknowledge.worldbank.org/handle/10986/16654

World Bank Group (2015). *Mauritius: Systematic Country Diagnostic, World Bank Group*. https://openknowledge.worldbank.org/handle/10986/23110?show=full

Wright, S. L., Thompson, R. C., & Galloway, T. S. (2013). The physical impacts of microplastics on marine organisms: A review. *Environmental Pollution*. https://doi.org/10.1016/j.envpol.2013.02.031

Yamashita, R., Hiki, N., Kashiwada, F., Takada, H., Mizukawa, K., Hardesty, B. D., Roman, L., Hyrenbach, D., Ryan, P. G., Dilley, B. J., Muñoz-Pérez, J. P., Valle, C. A., Pham, C. K., Frias, J., Nishizawa, B., Takahashi, A., Thiebot, J. -B., Will, A., Kokubun, N., ... Watanuki, Y. (2021). Plastic additives and legacy persistent organic pollutants in the preen gland oil of seabirds sampled across the globe. *Environmental Monitoring and Contaminants Research, 1*, 97–112. https://doi.org/10.5985/emcr.20210009

Zettler, E. R., Mincer, T. J., & Amaral-Zettler, L. A. (2013). Life in the "plastisphere": Microbial communities on plastic marine debris. *Environmental Science and Technology, 47*, 7137–7146. https://doi.org/10.1021/es401288x

Chapter 4
Legal and Policy Frameworks to Address Marine Litter Through Improved Livelihoods

Peter Manyara, Karen Raubenheimer, and Zaynab Sadan

Summary This chapter provides an overview of the international and regional legal and policy frameworks relevant to the prevention and management of marine litter. These instruments set the obligations and guidance for national action of participating countries. Legal and policy responses by governments provide an opportunity to address the many drivers of marine litter across the life cycle, from the design of products to the management of the waste they generate. Public awareness, consumer behaviour and industry engagement also play key roles in preventing marine litter. These interventions alone remain voluntary, fragmented and insufficient to tackle the marine litter problem. The national and/or regional responsibility of parties to prevent marine litter as established by these frameworks is not unique to the countries of Africa, and many of the barriers to effective compliance are shared with developing countries in other regions. The social context in which national implementation measures must operate can be unique to countries or regions. This chapter summarises the duties established by the legal and policy frameworks at the international and regional levels that may be applied to the issue of marine litter. It provides an African context to the barriers and drivers of effective implementation of national measures in compliance with international obligations. The scope of this chapter extends beyond the responsibility to prevent marine pollution, to establish a holistic and integrated duty of governments to provide a healthy environment and sustainable livelihoods as recognised in the global Sustainable Development Goals (SDGs). The review of these international, regional and national legal and policy frameworks therefore considers the inclusion of these broader principles to underpin prevention and management of marine litter.

P. Manyara
Marine Plastics and Coastal Communities (MARPLASTICCs), IUCN-International Union for Conservation of Nature, Eastern and Southern Africa Region, Nairobi, Kenya

K. Raubenheimer
Australian National Center for Ocean Resources and Security (ANCORS), University of Wollongong, Wollongong, Australia

Z. Sadan (✉)
WWF South Africa, Circular Plastics Economy Programme, Cape Town, South Africa
e-mail: zsadan@wwf.org.za

© The Author(s) 2023
T. Maes and F. Preston-Whyte (eds.), *The African Marine Litter Outlook*,
https://doi.org/10.1007/978-3-031-08626-7_4

Keywords Policy · SDGs · International · Regional · Africa · Policy and legal frameworks · Plastic pollution · Blue economy · Circular economy

4.1 Introduction

The development of effective policies and legal frameworks for the prevention and management of marine litter is driven by the responsibilities of governments to protect their environment, support the rights of their citizens, and, increasingly, public awareness of the consequences of the unsustainable generation of waste and inadequate management thereof. In addition, obligations are also set in legal and policy frameworks which aim to protect the global commons of the oceans and prevent transboundary harm to high seas marine environments and areas under the national jurisdiction of other coastal states.

The prevention of marine litter in African countries is hampered by challenges common to developing countries worldwide. These are areas to be considered for global and regional capacity building and funding opportunities to underpin short-, medium- and long-term solutions. Importantly, the design of legislative and/or policy response options must consider these challenges and integrate measures to remove or reduce their effect.

4.1.1 Poor, Inadequate and Fragmented Data, Information and Reporting

Across Africa, there are limited data on the quantities, composition and fate of marine litter (refer to Chap. 2 for more details). More broadly, there are similar limitations on data for rivers inputs and for waste management, particularly for rural areas (UNEP, 2018b). There are few exceptions, and some progress is being made. In South Africa, where research has been conducted over many years, beach litter has been tracked in a limited number of areas over time (Ryan, 2020). Kenya is also making progress in this regard. The need for establishing regular monitoring programmes, including beach litter surveys, has been well illustrated by the long-term data set developed in Seychelles, providing important data on litter fluxes to the islands (Dunlop et al., 2020).

4.1.2 A Lack of Targets and Metrics to Track Action and Progress Towards Goals

Without robust data, it is challenging to determine baselines from which targets can be set, including the development of indicators to provide the metrics to measure progress towards those targets. In the absence of such targets and metrics, it is possible to adopt regionally agreed targets and indicators that have been adopted and are appropriate to the local context. Some data may be extrapolated from the national reports transmitted on an annual basis by Parties to the Basel Convention; however, the quantitative data is often lacking and relates to waste generated and subject to transboundary movements overall rather than specifically relating only to marine litter. The 17 SDGs of the 2030 Agenda (UN, 2015) and associated indicators can also be adopted at the national level (see Annex 4.1: SDG targets and indicators relevant to preventing marine litter, livelihoods, and safe environment). In addition, the post-2020 Global Biodiversity Framework (CBD, 2021) includes targets and indicators relevant to the conservation of the marine environment, which the prevention of marine litter can help to achieve. In recognition of the lack of metrics, the United Nations Environment Programme (UNEP) and the International Union for Conservation of Nature (IUCN) developed the "National Guidance for Plastic Pollution Hotspotting and Shaping Action" (UNEP, 2020) to provide a common methodological framework that enables countries to prioritise interventions to abate plastic pollution—which makes up the largest volume of marine litter. The hotspotting methodology is being applied in Kenya, Mozambique, Tanzania and South Africa.

4.1.3 Limited Research into the Environmental and Social Impacts, Drivers and Solutions

The impacts of marine litter extend beyond environmental pollution and ecosystem degradation. Livelihoods, recreation and human health are also negatively impacted once plastics enter the marine environment. Research on human health impacts is ongoing, but socioeconomic studies are less common, particularly in Africa. Policy design must consider the social and economic effects of measures, particularly on vulnerable communities. In addition, understanding the drivers of marine litter, such as behaviour and access to services, will assist in the selection and design of cost-effective measures that target high-impact solutions and can scale nationally. In contrast, the banning of plastic products can lead to job losses (Godfrey, 2019), and careful consideration should be given to the broad spectrum of possible policy interventions to determine the most appropriate based on evidence.

4.1.4 Poor Compliance and Enforcement of Existing Legislation

A number of countries across Africa have adopted legislation or are in the process thereof, to ban or tax plastic bags of certain thickness, as well as a limited selection of single-use plastics (UNEP, 2018b). Poor enforcement of these measures has challenged their effectiveness (Oelofse and Godfrey, 2008; Jambeck et al., 2018; SADC, 2021). In addition, a lack of enforcement in one country can affect neighbouring countries. For example, the ban on plastic bags in Kenya has been undermined by the illegal trade in bags across land borders (Godfrey, 2019). Where economic incentives have been applied, such as the industry-led PETCO (the South African PET Plastic Recycling Company) system in South Africa, greater success has been achieved in the collection of plastic bottles and, therefore, a reduction in their contribution to marine litter. Despite these efforts, non-recyclable plastics and plastics of low value are still prevalent in the environment, prompting the South African government to move towards mandatory Extended Producer Responsibility schemes (see Box: EPR in the African context—the example of South Africa). It should be noted that without curtailing the escalating rates of production, the efficacy of even the most successful recovery schemes such as PETCO will continue to be challenged (Ryan et al., 2021a).

4.1.5   Absence of Integration of Environmental Justice in Waste Management

SDG 16 encourages the need for environmental justice through promoting peaceful and inclusive societies for sustainable development, providing access to justice for all and building effective, accountable and inclusive institutions at all levels (see Annex 4.1). This includes ensuring responsive, inclusive, participatory and representative decision-making at all levels (SDG Target 16.7) and ensuring the public has access to information (SDG Target 16.10). This applies to the degradation of ecosystem services and human harm caused by plastic pollution and associated chemicals, particularly the harm caused by the open burning of plastics (CIEL, 2019a, 2019b). Communities living in marginal lands near waste accumulation areas are more prone to water-borne and other diseases transmitted by pests and animals attracted to the waste (SADC, 2021). In some cases, the informal waste sector is dominated by women who are often exposed to unsafe working conditions. The livelihoods of this sector can also be reduced should formal waste systems be introduced (such as EPR schemes) that do not adequately incorporate the informal sector by reducing access to waste streams traditionally in their domain (US EPA, 2020). South Africa developed the Waste Picker Integration Guideline (DEFF, 2020b), which could be adapted to the local context in other African countries.

4.1.6 High Level of Product Importation not Matched by Appropriate Waste Management Capacity

Historically, there was limited production of plastic products in Africa. However, despite a continued reliance on imports, production is on the increase, fuelled by growing economies in the region (Africa Business, 2021). However, waste management services and capacity have not kept pace with increasing plastic consumption and the resulting waste generation (UNEP, 2018b). The Southern African Development Community (SADC) lists some of the drivers of unsustainable waste management as high volumes of waste generation coupled with poor waste management capacity, primarily due to limited adoption and high costs of appropriate disposal technologies and methods (SADC, 2021). In addition, electronic and electrical waste and other wastes imported into African countries have contributed to the pressure on already stressed and inadequate waste management services. Therefore, it is strongly recommended that, where recycling facilities are not readily available for products, African countries should avoid producing or importing those items (UNEP, 2020) or work more closely with producers to appropriately manage the product at the end of life, including promoting the principle of design for reuse and recycling.

4.1.7 Underfunded Waste Management Services and Limited Use of Market-Based Instruments

A lack of end markets for the reuse, recycling and recovery of wastes is a key contributor to poor waste services in Africa (UNEP, 2018b). Several market-based instruments have been implemented worldwide under differing socioeconomic contexts. These provide examples of best practices for incorporating the polluter pays principle through financial incentives (and disincentives). They can also promote the waste hierarchy, providing support for recycling (BRS, 2013, 2019a, 2019b; OECD, 2018).

Despite these limitations, waste pickers in South Africa have formed an association to promote and protect their livelihoods. The South African Waste Pickers Association (SAWPA) has met with waste pickers in Kenya to promote the establishment of a similar association and formally bring the various waste picker groups of Kenya together. These associations intend to advocate for good working conditions and gain recognition for waste picking as an essential service by decision-makers and other stakeholders. The waste picker groups recognise the need to strengthen this network across Africa further to "help fight false solutions that are presented in the waste sector" (GAIA, 2020).

4.2   The Role of Legal and Policy Frameworks
in the African Context and the Promotion of Equity

The international legal framework broadly provides for the prevention of marine pollution, thereby establishing the duty to protect the marine environment from plastic pollution. Only a handful of binding international agreements address the issue of marine litter directly or include measures that would directly contribute to the prevention thereof. The binding agreements include the Basel Convention (Convention on the Control of Transboundary Movements of Hazardous Wastes and Their Disposal, 1989), the Stockholm Convention (Convention on Persistent Organic Pollutants, 2001), MARPOL (International Convention for the Prevention of Pollution from Ships) Annex V (Regulations for the Prevention of Pollution by Garbage from Ships, 2011) and the London Convention (Convention on the Prevention of Marine Pollution by Dumping of Wastes and Other Matter, 1972). The United Nations (UN) Law of the Sea Convention mandates the prevention of marine pollution from all sources, including plastics, and applies to land-locked countries that contribute to marine plastic pollution via rivers and other pathways. The UN Fish Stocks Agreement requires the minimisation of wastes, discards and catches resulting from abandoned, lost, or otherwise discarded fishing gear (ALDFG), which is mostly made of plastics. Some voluntary global instruments target marine litter directly, such as SDG 14, which focuses on life below water and specifically targets a reduction of marine litter. Regional instruments have been adopted to protect the oceans in different regions, including those surrounding Africa. Implementation of international and regional instruments occurs at the national level, and the role of these instruments, both binding and voluntary, is to harmonise and facilitate national action. In some cases, initiatives taken by nongovernment institutions have provided a valuable basis for strengthening and the implementation of the legal and policy frameworks (see Sect. 4.4.3 on relevant regional and sub-regional marine litter initiatives). An approach across geographical scales is important; international strategies mobilise resources while increasing awareness on a global scale, however, awareness and will for action on a national level is essential to address the issue of marine litter. This is one of the key tracks but not the only one. An international treaty would allow to place the problem on the environmental agenda at the global level, mobilising resources, but without a awareness and will at the local level, the issue might remain a matter for experts preaching to the convinced.

The issue of marine litter raises several governance failures beyond the environmental degradation of marine ecosystems. Failure to protect the marine environment from marine litter damages the livelihoods of many who rely on these ecosystem services, particularly subsistence fishers. Thus, the right of individuals to a healthy and productive environment (Knox, 2020) is undermined. In addition, the right to decent work and economic growth (SDG8), sustainable cities and communities (SDG11) and, importantly for Africa, poverty eradication (SDG1) through job creation is denied. These factors are critical for security, peace and reducing displacement.

The existing legal and policy frameworks, both international and regional, present a fragmented approach to preventing and managing marine litter. The current frameworks are commonly assessed by evaluating those instruments that aim to prevent marine pollution, protect species and biodiversity or manage chemicals and waste (UNEP, 2017). In the absence of a global agreement to govern plastic pollution, the option to strengthen measures under the current framework and the coordination thereof towards a common goal remains a viable option to reduce marine litter. The elements of a new global agreement to address plastic pollution will be negotiated under the mandate of Resolution 5/14 adopted at the resumed fifth session of the United Nations Environment Assembly (UNEA). However, ignoring the co-benefits obtained from preventing marine litter presents a missed opportunity to address SDG8 on decent work and economic growth, SDG11 on making human settlements inclusive, safe, resilient and sustainable and SDG1 on ending poverty. Addressing these SDGs in the context of marine litter will move Africa towards achieving SDG12 on responsible consumption and production, thereby reducing the generation of wastes that may become marine litter.

The 17 SDGs are elaborated in relevant targets specific to achieving each goal. More than 250 indicators further support these targets for measuring progress towards each goal. Poverty reduction under SDG1 is particularly important in the African context. Poverty is a primary contributor to environmental degradation in developing countries (Masron & Subramaniam, 2019). A key component in achieving poverty reduction is economic growth (Ladan, 2018). Indicator 1.2.1 tracks progress towards halving the population living below the national poverty line. More specific to the drivers of marine litter, and in support of the right to a healthy environment and livelihoods, is target 11.6, aiming to reduce by 2030 the adverse per capita environmental impact of cities, by paying particular attention to municipal and other waste management, tracked through indicator 11.6.1 on the proportion of municipal solid waste collected and managed in controlled facilities. Target 12.5 aims to substantially reduce waste generation by 2030 through prevention, reduction, recycling and reuse, using indicator 12.5.1 to track national recycling rates and tonnes of material recycled.

Underpinning these duties by States is the need to develop domestic systems for sustainable financing to subsidise waste management systems and address leakage of litter into the marine environment. By incorporating the polluter pays principle, whereby producers must contribute financially and physically to the management of their products at the end of life, African countries can increase private sector investment in waste management services and create sustainable jobs. This will, in particular, benefit the informal waste sector (US EPA, 2020). Such policies can assist in measuring progress towards the goal of ensuring significant mobilisation of resources from a variety of sources to implement programmes and policies to end poverty in all its dimensions. This is tracked through SDG indicator 1.a.1 with a metric of the proportion of domestically generated resources allocated by the government directly to poverty reduction programmes.

By focusing on these targets and indicators, Africa can move towards achieving SDG14 on life below water, for which target 14.1 aims to prevent and significantly reduce marine pollution of all kinds, in particular from land-based activities, by 2025, including marine debris. Indicator 14.1.1 lists floating plastic debris density as a metric for this target. To address the challenges prevalent in Africa that exacerbate the drivers of marine litter, this chapter focuses on the national duties established under international and regional instruments to:

- Prevent and remove marine litter from land- and sea-based sources;
- Provide a healthy and productive environment;
- Provide sustainable economic growth that supports sustainable livelihoods with no environmental harm;
- Strengthen the science–policy interface through improved knowledge management relevant to Africa, support for scientific research and greater regional cooperation, including sharing of best practices and
- Establish mechanisms for sustainable financing of waste management, with a focus on job creation and geographic coverage of collection systems.

Integration of the above principles and duties to develop holistic waste management services using a source-to-sea approach (UNEP, 2021a, 2021b, 2021c) can assist countries in working towards a number of global priorities. These include the contribution of all phases of the plastics life cycle to climate change (CIEL, 2019a), to human health (CIEL, 2019b), the Post-2020 Global Biodiversity Framework towards the vision of living in harmony with nature (CBD, 2021), and the management of chemicals and waste beyond 2020 (SAICM, 2021). Together with the development of circular systems, integrating these approaches will improve the current fragmented policy frameworks and help deliver on the three global priority issues of climate, biodiversity and pollution (UNEP, 2021a, 2021b, 2021c).

4.3 International Legal and Policy Frameworks of Relevance to Marine Plastic Pollution in Africa

The global community has made significant strides in establishing international frameworks that support countries in addressing marine pollution challenges within and beyond their jurisdictions. In a period spanning 50 years, numerous international legal and policy instruments have been adopted, comprising conventions, or agreements, regulations, strategies, action plans, programmes, guidelines, etc. (see Annex 4.2). Some of the international instruments are legally binding to Parties that have expressed consent to be bound. In contrast others provide for voluntary participation or coordination and cooperation by States and other relevant actors. A number of studies, such as Lebreton and Andrady (2019), UNEP (2017) and Bergmann and others (2015), have highlighted the successes and limitations of the implementation of existing instruments in addressing marine litter. Some of the gaps identified include deficiencies in legislation or lack of implementation thereof, low cooperation and insufficient participation of States, inadequate data on marine litter and a fragmented framework. Recommendations arising from these studies towards addressing the gaps include exploring measures going beyond basic amendment to existing instruments and the need to develop an internationally legally binding agreement on addressing plastic pollution.

Many regions and States have incorporated into national law, as needed, some of the obligations enshrined in the international instruments and put in place measures to address marine litter. Yet, for many States or regional groups of States, ratified legal instruments such as treaties may be directly applicable and have primacy over national law, including before national courts, without having the necessity to adopt transposing national measures. Indeed, some regional instruments go further in addressing the gaps in international instruments by developing regional measures and action plans specific to marine litter (UNEP, 2017a, 2017b). For example, several African States have established measures such as prohibiting certain leakage-prone products (see Sect. 4.5). The shortcomings of the current global regulatory framework (see Sect. 4.2) were highlighted in the 2020 African Group contribution towards the UNEA 5 process[1]. There are instances where some frameworks allow a margin of discretion to Parties wishing to adopt more stringent measures.

The following subsections highlight some of the key legally binding international instruments that may support efforts by countries to address marine litter and plastic pollution within their scope of mandate and the various voluntary strategies, action plans, programmes and guidelines that have been adopted as support mechanisms.

The key legally binding instruments analysed in Sect. 4.3.1 are outlined in Table 4.1.

4.3.1   Global Conventions and Protocols

The various conventions and protocols outlined herein define the scope of the legally binding international governance framework for countries to strengthen mechanisms to protect the marine environment from human activities. Nearly all African countries have signed or ratified three global instruments: the UNCLOS, Basel and Stockholm conventions, but less than 30% have signed to the London, MARPOL Annex V and UN Watercourses Conventions (Table 4.1, details in Fig. 4.1a, b). Despite the domestication and implementation of the obligations of the international legal governance framework, recent assessments continue to highlight the worsening problem of marine plastic pollution and present an undesirable future outlook under a business-as-usual scenario where the total

---

[1] https://wedocs.unep.org/bitstream/handle/20.500.11822/34194/African%20Group%20Item%205.pdf?sequence=2&isAllowed=y.

Table 4.1 Number of African countries out of 54 that have ratified, accessioned or approved the various global instruments in this section (Data sources: United Nations Treaty Collection database; and convention websites)

Treaty name	Details of treaty	No. of contracting parties/members
UNCLOS	United Nations Convention on the Law of the Sea (UNCLOS, 1982)	47
London Convention	Convention on the Prevention of Marine Pollution by Dumping of Wastes and Other Matter (London Convention, 1972)	16
London Protocol	Protocol to the Convention on the Prevention of Marine Pollution by Dumping of Wastes and Other Matter, 1972 (London Protocol, 1996)	10
MARPOL 73/78 (Annex V)	International Convention for the Prevention of Pollution from Ships; Annex V—Regulations for the Prevention of Pollution by Garbage from Ships (MARPOL 73/78: Annex V, 2011)	36
Basel Convention	Convention on the Control of Transboundary Movements of Hazardous Wastes and Their Disposal (Basel Convention, 1989)	53
UNWC	Convention on the Law of the Non-Navigational Uses of International Watercourses (UN Watercourse Convention, 1997)	13
Stockholm Convention	Convention on Persistent Organic Pollutants (Stockholm Convention, 2001)	53
CBD	Convention on Biological Diversity (CBD 1992)	54

amount of plastic waste entering the ocean is projected to nearly triple by 2040 (Pew Charitable Trusts & SYSTEMIQ, 2020; Boucher et al., 2020; Williams & Rangel-Buitrago, 2019; Geyer et al., 2017). Global plastic waste generation is projected to triple from an estimated 60 to 100 million tonnes in 2015 to 155–265 tonnes by 2060, with Africa and Asia contributing disproportionately large shares of the total (Lebreton & Andrady, 2019). This points to the need for further scientific and policy exploration on alternative complementary or new instruments and platforms to effectively slow down plastic leakages into the environment and for increased efforts to strengthen implementation and compliance to existing instruments. Among the proposed policy responses to be debated through the UN Environment Assembly (UNEA) process is to establish a global treaty on plastics to reduce or eliminate the flow and leakage of related litter to the marine environment. A potential plastics agreement aims to address ongoing governance gaps and combat plastic pollution throughout its life-cycle stages. Three of its key goals could be to reduce virgin plastic production and consumption, facilitate for safe circularity of plastics and eliminate plastics pollution in the environment (Simon et al., 2021).

UNCLOS

The United Nations Convention on the Law of the Sea sets out the obligation for Parties to take all measures consistent with this Convention necessary to prevent, reduce and control pollution of the marine environment from any source (UNEP, 2017a, 2017b). The Convention addresses pollution from a number of sources, including land-based, seabed activities, dumping, from vessels and from or through the atmosphere. It prohibits dumping within the territorial sea and the exclusive economic zone or onto the continental shelf without the expression of prior approval of the coastal State. It mandates States to adopt the necessary laws and regulations and harmonise policies at the appropriate regional level. A recent study noted that many African States have not legislated their full maritime zone benefits available under UNCLOS nor defined the extent of their sovereign rights, obligations and jurisdictions through such legislation (Surbun, 2021). For example, Okonkwo (2017) notes that several unresolved maritime boundary disputes have slowed down the maritime boundary delimitation process, which seems not to be a priority in the absence of incursions by neighbouring countries. This context underlines the overall absence of political will to legislate the Convention's obligations fully and together, with lack of resources, undermines its full implementation by countries.

The Convention recognises pollution as the direct or indirect addition of substances or energy into the marine environment resulting in deleterious effects, including harm to living resources and marine life, hazards to human health, a hindrance to marine activities, including fishing, and other legitimate uses of the sea, impairment of quality for the use of seawater and reduction of amenities. Thus, at the national level, the implications of these provisions aim to safeguard the ecological benefits that the oceans provide, thereby supporting economic development. For example, South Africa's Ocean Economy is projected to contribute US $ 12 billion to the national GDP by 2033 and create over one million jobs (DEFF, 2017), and figures that rise if all African States are considered. Additional detail on the value of Africa's Blue Economy is provided in Chap. 1.

London Convention and Protocol

The 1972 London Convention and 1996 Protocol on the Prevention of Marine Pollution by Dumping of Wastes and Other Matter) promote the effective control of

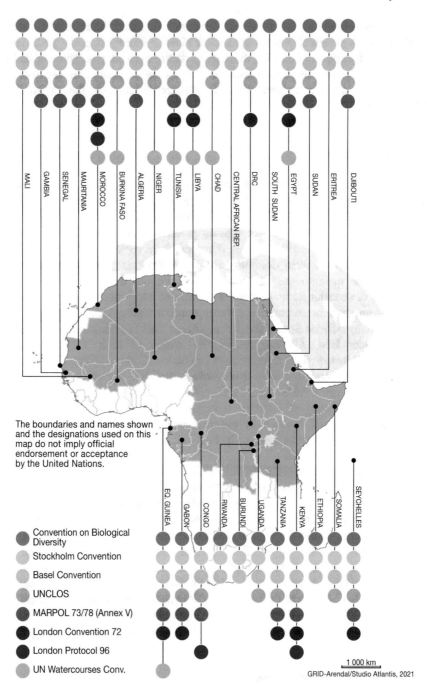

Fig. 4.1 African countries covered by relevant international conventions (ratified or accessioned)

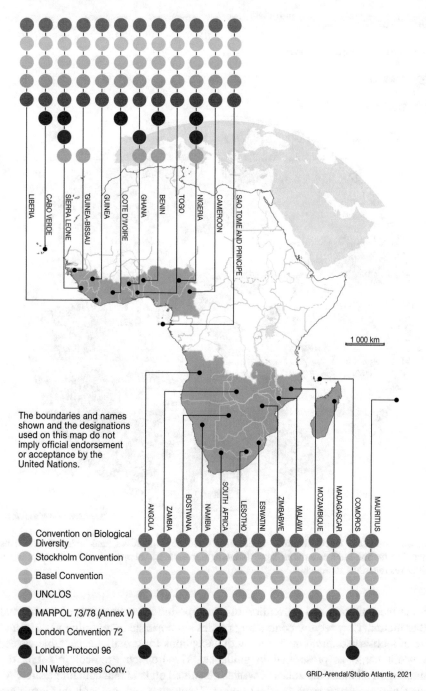

The boundaries and names shown and the designations used on this map do not imply official endorsement or acceptance by the United Nations.

GRID-Arendal/Studio Atlantis, 2021

Fig. 4.1 (continued)

Orientations

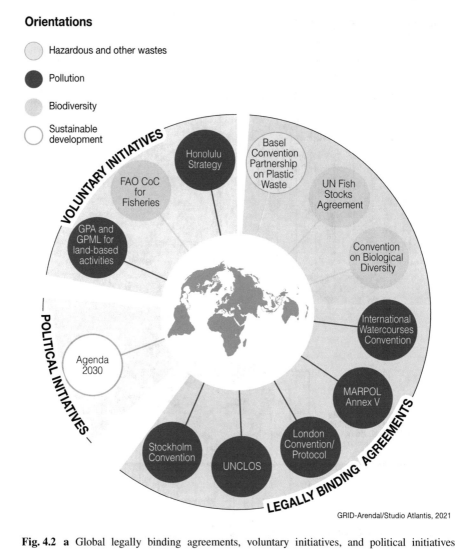

Fig. 4.2 a Global legally binding agreements, voluntary initiatives, and political initiatives regarding marine litter. **b** African legally binding agreements, voluntary initiatives and political initiatives regarding marine litter

all marine pollution and prevention of ocean pollution by dumping of wastes and other matter. They require contracting parties to cooperate in reporting vessels, and aircrafts observed dumping at sea, with exceptions for possibly acceptable wastes or spoilt cargo as prescribed in guidance. The London Protocol prohibits the dumping in all maritime zones of wastes generated on land that contain plastics. As a new and emerging issue for the London Convention and Protocol, its Governing bodies recommended action to combat marine litter through identification and control of marine litter at source, monitoring, additional studies and

Orientations

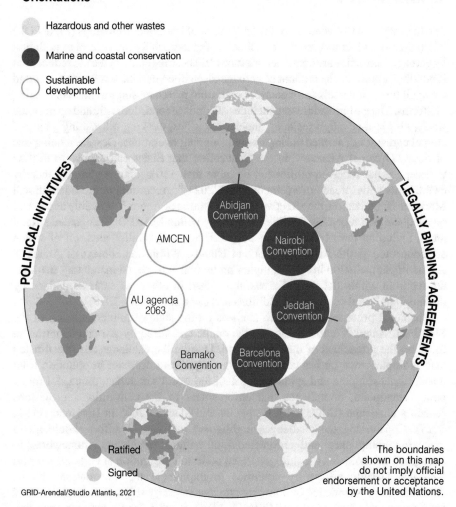

Fig. 4.2 (continued)

knowledge-sharing. As of 2019, 16 African countries were parties to the Convention and 10 to the Protocol (IMO, 2019a), with only six being Parties to both.

The International Maritime Organization (IMO) has outlined some benefits to countries that are party to the London Protocol.[2] Some of the benefits outlined include social, political and economical, such as those derived from a healthy marine environment or efficiencies in trade negotiations.

---

[2] Benefits of being a party to the London Protocol (https://wwwcdn.imo.org/localresources/en/Our Work/Environment/Documents/Benefits%20LP.pdf).

MARPOL Annex V

The International Convention for the Prevention of Pollution from Ships (MARPOL) 73/78 includes several Annexes, of which the regulations for the control of pollution by garbage from ships are contained in Annex V. These regulations seek to reduce, and if possible, eliminate the amount of garbage discharged into the sea from vessels and cover all types of vessels from pleasure crafts and merchant ships to fixed or floating platforms. The garbage addressed includes all kinds of food, domestic and operational waste, all plastics, cargo residues, incinerator ashes, cooking oil, fishing gear and animal carcasses generated during normal shipping operations. Despite it being one of three "Optional Annexes" in the Convention that States can choose to decline to accept, it has received ratification by more than 150 countries globally, thereby enabling its widespread enforcement (IMO, 2021)[3]. According to the International Maritime Organization, the effective implementation of Annex V is mainly dependent on adequate port reception facilities, especially within MARPOL designated special areas, including the Mediterranean Sea, Red Sea and "Gulfs" area, all of which are connected to the African States and Islands. Within the context of Annex V and its implementation in Africa, higher protection and special mandatory methods are required within the special areas, more than in other sea areas as defined by oceanographic and ecological conditions and sea traffic (IMO, 2019)[4].

Chapter 2 provided an in-depth analysis on the state of implementation of the MARPOL Annex V and expands on the challenges faced by African countries in fulfilling the provisions of the Convention. Despite the challenges, it is estimated that its implementation has contributed to a significant decrease in pollution from international shipping and applies to 99% of the world's merchant tonnage. From a policy standpoint, compliance and enforcement of the convention's provisions remain a challenge (IMO, 2019; Carpenter and MacGill 2005 in Bergmann et al., 2015). Considering the challenges highlighted in Chap. 2, effort should go to reviewing compliance and enforcement of national regulatory frameworks to strengthen mechanisms for ship-generated pollution prevention at harbour and port facilities. Despite efforts by countries to implement the Convention and its Annexes, there is evidence of dumping garbage at sea by ships, in contravention of the Convention (Ryan et al., 2019; IMO, 2018). Results from multiyear recording and monitoring of accumulation of plastic debris along Inaccessible Island, a remote, uninhabited island in the central South Atlantic Ocean, show increased plastic accumulation growth in the recent past. The research found that 90% of plastic bottles stranded on the island had been manufactured recently, with more than 83% of new bottles of Asian origin. The time-since-manufacture stamps on the bottles are significant in assigning the likely responsible source from ships (Ryan et al., 2019). Similar evidence of illegal dumping of plastics from ships

---

[3] Prevention of Pollution by Garbage from Ships (https://www.imo.org/en/OurWork/Environment/Pages/Garbage-Default.aspx).

[4] Special Areas under MARPOL (https://www.imo.org/en/OurWork/Environment/Pages/Special-Areas-Marpol.aspx).

operating in Asia has been observed through scrutiny of plastic bottles found on beaches in Kenya and South Africa (Ryan, 2020; Ryan et al., 2021a). These sea-based and transboundary sources of plastic bottles are found on all beaches, though their signal is diluted close to local land-based sources.

There are benefits derived when countries implement MARPOL Annex V. For instance, Nigeria, Africa's largest economy and host to a high level of shipping traffic, domesticated the MARPOL Convention as part of its port reforms starting in 2000. The reforms empowered the Nigerian Maritime Administration and Safety Agency (NIMASA) as an oversight body and created an enabling environment for public–private investments in port infrastructures. A private waste management agency secured a long-term contract to manage port reception facilities in Nigeria's six largest ports (UNEP, 2017b). Such legislative and institutional reforms boosted confidence in the private sector to invest more than 70 million US dollars towards shipping waste management infrastructure (Obi, 2009). The contribution of marine litter in Africa from transboundary and sea-based sources (Ryan, 2020; Ryan et al., 2021a) has further economic repercussions, as plastic accounts for 94–98% of all litter on Cape Town beaches, where the tourism sector directly employed about 44 thousand people (Takunda & Blottnitz, 2019; City of Cape Town, 2019). A recent study by Jain et al. (2021) estimated that plastic litter on the beaches of Cape Town could lead to losses of up to R 8.5 billion in total coastal tourism revenue, representing 91% of total coastal tourism revenue and 67% of overall tourism revenue.

Basel Convention

The 1989 Basel Convention on Control of Transboundary Movements of Hazardous Wastes and Their Disposal was established due to increasing public outcry in Africa and other parts of the developing world, in relation to deposits of toxic wastes imported from abroad. The Convention aims to reduce hazardous and other waste generation, promote environmentally sound management, restrict transboundary movements of hazardous and other wastes and provide a regulatory system on where transboundary movements are permissible geographically.

In 2019, the Conference of Parties to the Convention adopted amendments to three annexes to enhance the control of transboundary movements of plastic waste and to clarify the convention's scope as it applies to such waste. The amendments specified the types of plastic wastes that are presumed to be hazardous or not, and as such, which would be subject to the Prior Informed Consent (PIC) procedure. Non-hazardous waste in the new entry to the annex is understood to include mixtures of plastic wastes consisting of commonly used plastics, such as polyethylene (PE), polypropylene (PP) or polyethylene terephthalate (PET), provided they are destined for separate recycling, in an environmentally sound manner, and are almost free from contamination.

These amendments may have several ramifications, especially as it relates to waste traded between developed and developing countries. Overall, they are intended to strengthen the regulation of plastic waste shipped to countries that cannot manage it in an environmentally sound manner. Africa is already affected by illegal traffic and cross-border movements of wastes from countries and companies seeking cheaper and less regulated disposal options for their waste (Mail & Guardian, 2020; Schluep et al. 2012). The dumping of hazardous wastes into Africa has occurred for decades, such as the dumping of 500 tonnes of toxic waste near Abidjan, Cote d'Ivoire, in 2006, which killed around 15 people and left thousands more experiencing severe illness (Mail & Guardian, 2020). Much of this waste is treated using environmentally unsound practices, further exposing the population to toxic compounds (BRS, 2019a, 2019b, 2019c). For example, a study at a Ghanaian processing site recorded amongst the highest exposure levels of POPs due to unsound management of imported plastic wastes (Bruce-Vanderpuije et al., 2019).

The 2018 Africa Waste Management Outlook highlighted, as a major weakness, the lack of data on transboundary movements of hazardous waste across many African countries and the lack of transmission by Parties of their annual national reports as required by the Convention (UNEP, 2018a, 2018b, 2018c). A number of waste-exporting countries now face waste management challenges due to the recent Basel amendments, resulting in further pressure to find new end markets for their wastes. Recent news publications have reported on major oil companies lobbying the United States government in their ongoing trade negotiations to pressure African Countries to ease their stance against plastic waste imports (Guardian, 2020; New York Times, 2020). Note that the United States is a Signatory to the Convention but has not ratified it. As much as compliance with the amended provisions of the Basel Convention provides for reduced inflows of hazardous wastes to African countries, it needs to be accompanied by strengthened mechanisms to implement and enforce the Convention at the national level.

The Basel amendments came into effect on the 1st of January 2021 for all Parties that had not submitted a notification of non-acceptance of the amendments. As none of the African states that are Parties to the Basel Convention submitted such notification, the amendments entered into force for all the 53 Parties across Africa. The categorisation as per the amendments of specific PE, PP and PET waste as non-hazardous is relevant, as these are some of the most recycled polymers across the continent (UNEP, 2018a, 2018b, 2018c). While it is predicted that exports of well-sorted wastes may continue uninterrupted. Though subject to additional inspection, African countries may face challenges in exporting mixed-plastic bales elsewhere due to the increased administration of their movement due to new measures adopted to implement the amendments (Resource Recycling, 2019). Such challenges may present opportunities for African countries to invest in expanding environmentally sound domestic recycling infrastructure and services, thereby benefitting informal sector actors who recover and supply such post-consumer recyclables to the recycling economy. The need for Africa to optimise the benefits provided by the informal sector in managing plastic waste and leakages through positive engagement, support and integration has been highlighted (UNEP, 2018b).

UN Watercourses Convention

The UN Watercourses Convention 1997 (Convention on the Law of the Non-navigational Uses of International Watercourses) requires Parties that use international watercourses to take appropriate measures to prevent harm to other watercourse Parties, including preventing, reducing and controlling pollution, which includes plastics. In an analysis undertaken by UNEP as part of its UNEA process, this convention could cover a broader inland scope of application to sources and activities. As one of the most recent frameworks, it has a low level of government participation. Out of the 40 Parties who have approved, accepted or ratified the Convention, 13 are from Africa, with only four Signatories, namely Côte d'Ivoire, Namibia, South Africa and Tunisia.

Stockholm Convention

The Stockholm Convention 2001 (Convention on Persistent Organic Pollutants) provides measures to reduce or eliminate releases of persistent organic pollutants (POPs) as these present risks to human health and the environment, including marine ecosystems. It provides for the protection of human health and the environment by setting out the obligations for Parties to restrict or eliminate intentional production and use of chemicals listed under its annexes A and B; measures to control trade (imports and exports) in such chemicals; measures to reduce or eliminate releases of POPs from stockpiles and wastes; measures to reduce or eliminate release from unintentional production of chemicals listed in its Annex C. These measures encompass, among other things, production, use and disposal of additives that are POPs (e.g. PBDEs, PFOS, PFOA and SCCPs) in plastic products or during the manufacture of plastics. This is relevant for plastics recycling and reusing articles in use containing quantities of such regulated POPs. Open burning and incomplete incineration of plastic waste unintentionally produce and release of POPs. For example, recent assessments estimated the air pollution by noxious chemical gases from open burning of mismanaged plastic in 2018 as 233 kti in Kenya, 80 kt in Mozambique, 514 kt in South Africa and 129 kt in Tanzania (IUCN-EA-QUANTIS, 2020a, 2020b, 2020c, 2020d). The studies called for further investigations on open burning practices, backed up by field studies to accurately estimate the amount of mismanaged plastic waste burned across African countries.

POPs in plastics are relevant as they pose potential toxic risks (refer to Chap. 1), added to that posed by the plastic particles and leaching of plastic additives (Iñiguez et al., 2017). In general, additives to plastics constrain their recovery, recyclability and disposal, thereby reducing their contribution to a circular economy (Wagner & Schlummer, 2020). Therefore, the risks posed by POPs and related harmful plastics additives need to be addressed; otherwise, they inhibit economic development through job creation in the recycling sector. Recycling remains a prime investment area in Africa, where mismanaged waste plastics constitute a high proportion of generated plastic waste, and overall only 4% of plastic waste is recycled (Babayemi, 2019; UNEP, 2018a, 2018b, 2018c).

Convention on Biological Diversity (CBD)

The Convention on Biological Diversity is dedicated to promoting sustainable development. The Convention recognises that biological diversity is more than just plants, animals, microorganisms and their ecosystems and that it integrates people and ecosystem-derived needs. This Convention largely focuses on conservation of biological diversity with marine plastic litter and microplastics included in its targets (UNEP, 2017a, 2017b), specifically Aichi Biodiversity Targets 8 on reducing pollution pressures to biodiversity and Target 10 on minimisation of anthropogenic pressures. Parties to the Convention have also adopted key decisions, such as Decision COP XIII/10 which provides voluntary guidance for preventing and mitigating the impacts of marine litter. The decision also calls for the strengthening of existing legal frameworks to eliminate the production of microplastics and address wastes from fisheries and aquaculture (UNEP, 2017a, 2017b).

Towards preparations for the post-2020 global biodiversity framework, the Convention received a number of submissions that proposed for the Aichi Biodiversity Targets to serve as a basis for the targets in the post-2020 global biodiversity framework, but with modifications. Some of the feedback was to consider formulating new targets such as plastics and sustainable production, issues not already covered by the existing Targets (CBD, 2019). The revision of the targets could be relevant, given that its success in addressing marine litter remains debatable.

The Convention on Biological Diversity (CBD) is considered a universally accepted Convention, given its ratification by 196 Parties, with the United States being a Signatory only, but not yet a Party. All 54 African countries have ratified and are Parties to the Convention.[5]

4.3.2   *International Arrangements, Processes and Initiatives*

In addition to the above global legally binding treaties or conventions and their related protocols, there are complimentary global intergovernmental cooperative mechanisms, processes and strategies relevant in this regard. A number of African countries (Fig. 4.2a, b) engage with these global arrangements at the regional and national levels through various cooperative mechanisms. A few mechanisms of relevance in this context are highlighted in subsections below.

---

[5] List of parties to the convention on biological diversity (https://www.cbd.int/information/parties. shtml).

In domesticating their obligations as enshrined in the international instruments, African countries benefit from strengthened regional and national institutional, legal and policy frameworks, including creating avenues for sharing knowledge, technologies and funding to fulfil their related obligations in relation to preventing and addressing pollution.

Cumulatively, the global instruments encapsulate articles and provisions to address land- and sea-based sources of marine litter, call for and at times facilitate for development of regional and national strategies for pollution and waste prevention, address the potential effects of marine litter and plastic pollution to human health and the marine environment and contribute to fostering a healthy development of the world economy while considering the interests and needs of developing countries.

In addition, these global instruments support the advancement of scientific knowledge and facilitate information-sharing to better support regional and national decision-making. This includes supporting the development of national and regional marine science, technology centres, strengthening national scientific and technical research capabilities, collaboration in scientific and technical research and promoting access and exchange of data and analyses. The instruments also promote the sustainable financing of upstream and downstream interventions such as training and transferring environmentally sound technologies and know-how towards achieving their objectives. They also broadly point to the need for technical assistance to enable developing countries and countries with economies in transition to incorporate the goals and objectives of the various instruments into their national responsibilities to achieve their full implementation. While acknowledging the complexity in addressing marine litter, the instruments underline the need for strengthening efforts further upstream to change production and consumption patterns reduce the generation of waste and manage it.

Global Programme of Action for the Protection of the Marine Environment from Land-Based Activities/Global Partnership on Marine Litter (GPA/GPML)

The Global Partnership on Marine Litter was initiated under the Global Programme of Action for Protection of the Marine Environment from Land-based Activities (GPA). The GPA is hosted by the United Nations Environment Programme (UNEP) and brings together international agencies, governments, nongovernmental organisations, academia, the private sector, civil society and individuals to reduce the impacts of marine litter on economies, ecosystems and human health. Under it, the Global Partnership on Marine Litter (GPML) has been established to address the global challenge posed by marine litter, including plastics, by engaging all public and private sector stakeholders in the lifecycle of plastics and encouraging solutions by all sectors. The Global Partnership on Waste Management also has a complimentary process, which includes marine litter as one of nine focal areas.

Basel Convention Partnership on Plastic Waste (PWP)[6]

The Conference of the Parties to the Basel Convention established Plastic Waste Partnership (PWP) in 2019. The PWP aims to mobilise business, government, academic and civil society resources, interests and expertise, to improve and promote the environmentally sound management (ESM) of plastic waste at the global, regional and national levels and to prevent and minimise its generation. The activities of the PWP will contribute to efforts to reduce and eliminate the discharge of plastic waste and microplastics into the environment in general and, in particular, the marine environment. The PWP has established four project groups:

1. Plastic waste prevention and minimisation;
2. Plastic waste collection, recycling and other recovery including financing and related markets;
3. Transboundary movements of plastic waste and
4. Outreach, education and awareness-raising.

Eight pilot projects from the African region are among the initial 23 supported by the PWP during its inaugural phase. The pilot projects aim to improve and promote the environmentally sound management of plastic waste and prevent and minimise its generation. In general, the partnership operationalises the newly expanded mandate of the Basel Convention to address plastic pollution.

The efforts within the PWP are undertaken within the context of international cooperation activities by the Secretariat of the Basel, Rotterdam and Stockholm Conventions that amplify and ensure consistency between initiatives at the global level to ensure the ESM of plastic waste. This includes cooperation with entities such as the World Customs Organisation, the International Maritime Organisation and the World Trade Organisation.

United Nations Environment Assembly/Ad-Hoc Open-Ended Expert Group on Marine Litter and Microplastics (UNEA/AHEG): Africa Group

The United Nations Environment Assembly's Bureau of the Ad Hoc Open Ended Expert Group on Marine Litter and Microplastics (AHEG) has facilitated regional consultative structures to maximise transparency, inclusivity and participation. The Africa Group engages countries within the continent, facilitates consultations and feedback between the Chair and Bureau with the region and shares information on the progress and ongoing AHEG work. During the fourth Meeting of the AHEG, held online in November 2020, the Africa Group signalled an interest to explore the option of a new legally binding agreement with a shared vision to eliminate all discharge of plastic into the environment. Importantly, they reiterated the need for strengthening

---

[6] http://www.basel.int/Implementation/Plasticwaste/PlasticWastePartnership/tabid/8096/Default.aspx.

the means of implementation for related action, including through adequate and sustainable financial support, transfer of technology and capacity building.[7]

Small Island Developing States Accelerated Modalities of Action (Samoa Pathway)

The Samoa Pathway supports action to address marine pollution by developing effective partnerships, including developing and implementing of relevant arrangements and instruments. Acknowledging the waste management limitations in Small Island Developing States (SIDS), the Pathway calls for enhancing technical cooperation through the various Conventions and Protocols; the strengthening of national, regional and international mechanisms for managing waste, including marine plastic litter. Africa has six island states: Cape Verde, Comoros, Madagascar, Mauritius, São Tomé and Príncipe and Seychelles.

The Samoa Pathway covers the conservation of the marine environment through several Articles, with Article 58(b) addressing the marine plastic waste problem, and articles 70 and 71 covering the management of chemicals and waste, including hazardous waste, and remedial approaches such as reduction, reuse, recycling, recovery and return policies. Pathway implementation is reviewed through a high-level process at the UN General Assembly (UNGA) and national and regional levels.

As part of addressing marine litter, the 2019 Mid-Term Review of the SAMOA Pathway High Level Political Declaration called for addressing different types of waste through innovative approaches, including inter alia mismanaged plastic waste, chemical waste and marine litter, including plastic litter and microplastics.[8]

Group of Twenty (G20) and the Group of Seven (G7)

The Group of Twenty (G20) is a forum for international economic cooperation. It brings together the world's major economies to discuss global economics and finance issues. The G20 Environment Ministers adopted the "G20 Implementation Framework for Actions on Marine Plastic Litter", to tackle marine plastic litter and microplastics and their adverse impacts. South Africa is the only African country that is a member of the G20.

The Group of Seven (G7) is an intergovernmental organisation of the world's largest developed economies that meet periodically to address international economic and monetary issues. Through the Ocean Plastics Charter, the G7 members have articulated their commitment to take action towards a resource-efficient lifecycle management approach to plastics in the economy.

---

[7] https://wedocs.unep.org/bitstream/handle/20.500.11822/34194/African%20Group%20Item%205.pdf?sequence=2&isAllowed=y.

[8] https://www.un.org/pga/73/wp-content/uploads/sites/53/2019/08/SAMOA-MTR-FINAL.pdf.

Through the Charter, the G7 reiterates its commitment to mobilise and support the collaborative government, industry, academia, citizen and youth-led initiatives, including accelerating catalysing investments to address marine litter in global hotspots and vulnerable areas through public–private funding and capacity development, innovative solutions and coastal clean-up. It has advanced the G7 Plastics Innovation Challenge to address marine plastic litter by stimulating innovations, awareness-raising and improvements to the management of plastic, especially plastic waste, in developing countries.

At the G7 leader's summit held in Cornwall, UK, in June 2021, the leaders adopted the G7 2030 Nature Compact that commits to taking action to tackle increasing levels of plastic pollution in the ocean, including working through the UN Environment Assembly (UNEA) on options including strengthening existing instruments and a potential new agreement to address marine plastic litter.[9] The G20 and G7 are highlighted here, partly due to their power to influence global policy processes. Their membership comprises some of the wealthiest countries in the world that could partner with African countries to address marine litter. It is worth noting that the major plastics and chemicals producers are located across this group of countries, hence their relevance in contributing to addressing the plastics problem.

Honolulu Strategy: A Global Framework for Prevention and Management of Marine Debris[10]

The Honolulu Strategy is a framework for a global effort to reduce the ecological, human health and economic impacts of marine debris globally. It is designed to be used as a planning tool, reference framework and monitoring progress across programmes. The strategy elaborates a framework for collaboration and coordination among the multitude of stakeholders across the globe concerned with marine debris. It encourages participation and support on global, regional, national and local levels. The strategy encourages the establishment of appropriate mechanisms to facilitate the removal of marine debris. As a global strategy not directly addressing the unique issues, cultures and contexts across the 54 African countries, it has guided the development of regional and sub-regional strategies to address marine litter in Africa. Two recent examples include the "*Strategy for*

---

[9] https://www.whitehouse.gov/briefing-room/statements-releases/2021/06/13/carbis-bay-g7-sum mit-communique/.

[10] https://wedocs.unep.org/bitstream/handle/20.500.11822/10670/Honolulu%20strategy.pdf?seq uence=1&isAllowed=y.

Marine Waste: Guide to Action for Africa'[11] and the '*Western Indian Ocean Regional Action Plan on Marine Litter*".[12]

The strategy comprises three goals and associated strategies to reduce the amount and impact of marine debris from land- and sea-based sources and marine debris accumulations and aims for:

- reduced amounts and impacts of land-based sources of marine debris introduced into the sea (Goal A),
- reduced amounts and impacts of sea-based sources of marine debris, including solid waste; lost cargo; ALDFG; and abandoned vessels (Goal B) and
- reduced amounts and impacts of accumulated marine debris on shorelines, benthic habitats and pelagic waters introduced into the sea (Goal C).

The strategy provides countries with an opportunity to translate some of the strategies under each goal into concrete national policies and programmes aimed at preventing or reducing marine litter and plastic pollution into the marine environment.

UN Clean Seas Campaign[13]

The Clean Seas campaign is an initiative established to contribute to the objectives of UNEP's GPML. It serves as a platform to connect and rally governments, industry, civil society groups and individuals to be catalysts of change. It advocates for a transformation in habits, practices, standards and policies around the globe to dramatically reduce marine litter and its negative impacts. Since its launch in 2017, more than 62 countries have joined, making the Clean Seas Campaign the biggest, most powerful global coalition devoted to end marine plastic pollution. So far, the only African governments which have joined the campaign are Kenya and Rwanda.

IMO's Action Plan[14]

In 2018, the IMO's Marine Environment Protection Committee adopted an action plan to address marine plastic litter from ships through several measures to reduce plastic litter in the marine environment. The action plan, which runs until 2025, is established to enhance existing policy and regulatory frameworks and provides the IMO with a mechanism to identify specific outcomes and actions and introduce

---

[11] Strategy for Marine Waste: Guide to Action for Africa. https://sst.org.za/wp-content/uploads/2020/07/Marine-Waste-Strategy-Guide-to-Action-for-Africa.pdf.

[12] Western Indian Ocean Regional Action Plan on Marine Litter. https://nairobiconvention.org/Meeting%20Documents/December%202018/WIO-RAPMaLi_Full%20Revised%20Draft_2910 2018_Final.pdf.

[13] https://www.cleanseas.org/about.

[14] https://www.imo.org/en/MediaCentre/PressBriefings/Pages/20-marinelitteractionmecp73.aspx.

supportive measures to address the issue of marine plastic litter from ships. Some of the identified measures include studies on marine plastic litter from ships, adequacy of port reception facilities, fishing gear-related litter sources and awareness building on the impact of marine plastic litter. The IMO GloLitter Partnerships project will aid developing countries to prevent and reduce marine litter, with a focus on plastic, by identifying ways for the reduction of plastic applications in the maritime transport and fisheries industry.

4.4 Regional and Sub-Regional Frameworks and Initiatives

The regional and sub-regional frameworks covered in this section are reviewed through five lenses that are essential for success in an African context: (i) prevention or removal of marine litter, (ii) providing a healthy and productive environment, (iii) providing sustainable development through sustained livelihoods, decent work and economic growth, (iv) cooperate in science–policy interface towards improved knowledge management relevant to Africa and (v) sustainable financing of waste management.

4.4.1 Regional Frameworks and Initiatives

African Union

The African Union (AU) is a key driver for successful economic development in Africa. It is a continent-wide forum for Heads of State of Member States, or their representatives, to adopt coordinated positions.

African Union Blue Economy Strategy

The recent African Union Blue Economy Strategy (AU-IBAR, 2019) is the first of the AU's strategies or plans emphasising the economic development potential of marine (and freshwater) resources. It recognises pollution by chemicals and plastics as a key threat to the blue economy, including fisheries, aquaculture and tourism. See Chap. 1 for details on the blue economy and potential growth in Africa.

Agenda 2063 of the African Union

Agenda 2063, adopted in 2015 by the African Union, is a vision of how Africa aims to deliver inclusive and sustainable development. None of the flagship programmes to be undertaken during its first 10-year implementation plan

(2014–2023) are directly related to marine and coastal protection and management, nor the Ocean/Blue Economy. Still, the sixth goal states that a Blue Economy will accelerate economic growth. The priority areas for that goal are identified as marine resources and energy, and port operations and marine transport (AU, 2015). The seventh goal also addresses the blue economy by including priority areas such as sustainable resource management and biodiversity conservation and sustainable consumption and production patterns. These areas indirectly support the prevention and removal of marine litter. The implementation plan also calls for improved sanitation in cities, aiming for a minimum recycling rate of 50% of the waste generation and further supported by the roll-out of policies to enable the growth of urban waste recycling industries.

2050 Africa's Integrated Maritime Strategy

The 2050 Africa's Integrated Maritime (AIM) Strategy, developed by the AU in 2012, provides a framework for the region's protection and sustainable exploitation of Africa's Maritime Domain (AMD) for economic growth. One of the twelve strategic objectives include "protection of populations, including AMD heritage, assets, and critical infrastructure from maritime pollution and dumping of toxic and nuclear waste" (AU, 2012, pp. 12). This objective addresses explicitly the prevention and reduction of marine pollution via dumping from ships and maritime activities through maritime governance actions.

African Ministerial Conference on the Environment

The African Ministerial Conference on the Environment (AMCEN) was established in 1985 to provide advocacy for environmental protection in Africa, social and economic development and sustainment of basic human needs, including food security (UNEP, n.d.). This forum is represented by the Member States' Ministers of Environment and their representatives to adopt coordinated positions on various topics related to environmental issues, including but not limited to climate change, Blue Economy, the circular economy, biodiversity loss and plastic pollution.

At the 17th session of AMCEN in 2019, plastic pollution and marine litter prevention was addressed through the Ministerial Durban declaration themes of the circular economy, the Blue Economy and plastic pollution. Its decision 17/1, III (paragraph 11)[15] highlights and promotes the circular economy as a comprehensive approach to address plastic pollution. The key policy messages in the appendix to the Durban Declaration on environmental sustainability and prosperity in Africa (Section VIII) specifically highlight plastic pollution as a focus area. Here,

---

[15] African Ministerial Conference on the Environment Seventeenth session, November 2019. Report of the ministerial segment. AMCEN/17/9. Retrieved from: https://wedocs.unep.org/bitstream/han dle/20.500.11822/30786/AMCEN_17L1.pdf?sequence=1&isAllowed=y. Accessed: 11 June 2021.

continent-wide support for global action to address plastic pollution "to engage more effectively on global governance matters relating to plastic pollution" is expressed (p. 8). It further mentions that both the option of a new global agreement and reinforcement of existing agreements should be considered.

Bamako Convention

The Bamako Convention on the Ban of the Import into Africa and the Control of Transboundary Movement and Management of Hazardous Wastes within Africa (Bamako Convention) came into force in 1998. It is a treaty of African Member States prohibiting importing to Africa and ocean and inland water dumping or incineration of any hazardous (including radioactive) waste. The Bamako Convention responds to Article 11[16] of the Basel Convention, which encourages regional agreements on hazardous waste to help achieve its objectives. The Bamako Convention was also driven by incidents of hazardous waste being dumped in African countries by developed countries, causing major health and environmental impacts.

Only 29 African countries are Parties to the Bamako Convention.[17] South Africa and Nigeria have not ratified the Convention given the risk that it may inhibit their recycling economies, which involve transboundary trade of goods such as e-waste and plastic waste (Parliamentary Monitoring Group, 2008, 2014). Other possible concerns for states who are not party to the Convention may include the high financial investments required for effective implementation and the need for dedicated, skilled personnel (Ouguergouz, 1993).

Interestingly, the Bamako Convention responds to the increasing number of policy developments to address plastic pollution in its third Conference of the Parties (COP) in 2020. The COP-3 decision CB.3/8[18] contains multiple interventions to prevent plastic pollution in Africa, including to:

- invite Parties and other African countries to prohibit the manufacture, importation, sale and use of plastic bags and single-use plastic items;
- urge Parties to add "all forms of plastic wastes" to its Annex I—categories of waste which are considered hazardous, a more extensive list than that of the amendments to the Basel Convention in 2019;

---

[16] Basel Convention on the Control of Transboundary Movements of Hazardous Wastes and Their Disposal. Retrieved from: http://archive.basel.int/text/17Jun2010-conv-e.pdf. Accessed: 11 June 2021.

[17] The Parties to the Bamako Convention (as of February 2020) are: Angola, Benin, Burkina Faso, Burundi, Cameroon, Chad, Comoros, Congo, Côte d'Ivoire, Democratic Republic of Congo, Egypt, Ethiopia, Gabon, Gambia, Guinea-Bissau, Liberia, Libya, Mali, Mauritius, Mozambique, Niger, Rwanda, Senegal, Sudan, Togo, Tunisia, Uganda, United Republic of Tanzania and Zimbabwe.

[18] Conference of the Parties to the Bamako Convention on the Ban of the Import into Africa and the Control of Transboundary Movement and Management of Hazardous Wastes within Africa, Third meeting, February 2020. Meeting report. UNEP/BC/COP.3/11. Retrieved from: https://wedocs.unep.org/bitstream/handle/20.500.11822/32131/BamakoCOP3Report.pdf. Accessed: 11 June 2021.

- encourage Parties to participate in activities under the Basel Convention and, in particular, in the Partnership on Plastic Waste;
- increase awareness and education on the environmental and human health implications of plastic pollution and
- call for a new global legally binding agreement to address plastic pollution using a life cycle approach.

While this decision focuses on plastic pollution more broadly, it may indirectly help reduce and prevent some instances of plastic pollution and its contribution to marine litter.

Regional and Coordinating Centres Under the Basel and Stockholm Conventions

The Basel and Stockholm conventions benefit from a network of 23 Regional and Coordinating Centres for Capacity Building and Technology Transfer (BCRCs-SCRCs), six of which are located in Africa. These centres have been established to assist Parties to these conventions in implementing their obligations under these conventions, which includes promoting environmentally sound management of hazardous waste (including those containing or contaminated with POPs) and other waste. These centres took the initiative of establishing a working group on marine litter. They prepared a report on the challenges and measures to tackle marine litter plastics and microplastics and their POPs and EDC components[19] in 2016. Further, in 2017 and again in 2019, the conferences of parties to the Basel and Stockholm conventions mandated the regional centres to work on the impact of plastic waste, marine plastic litter, microplastics and measures for prevention and environmentally sound management.[20,21] A range of projects on plastic waste is being undertaken under the Basel and Stockholm Conventions' Regional Centre Small Grants Programme (SGP), known as "SGP on plastic waste". These projects are implemented by Basel Convention regional and coordinating centres and Stockholm Convention regional and sub-regional centres, funded by the Norwegian Agency for Development Cooperation (Norad). The projects aim to improve the management of plastic waste in partner countries and thus contribute towards preventing and significantly reducing marine pollution. In total, 15 projects have been selected for funding. They are being implemented in 2021–2022 in 32 beneficiary countries, 11 are countries from the African region.[22]

---

[19] UNEP/CHW.13/INF/29/Rev.1-UNEP/POPS/COP.8/INF/26/Rev.1.

[20] BC-13/11 (paragraph 14); BC-14/18 (paragraph 15) and SC-8/15 (paragraph 12); SC-9/14 (paragraph 14).

[21] UNEP/CHW.14/INF/29/Add.1 and UNEP/POPS/COP.9/INF/28/Add.1 (http://www.brsmeas. org/tabid/7832/language/en-US/Default.aspx).

[22] http://www.basel.int/Implementation/Plasticwaste/Technicalassistance/tabid/8402/Default. aspx.

Regional Fisheries Bodies

Regional fisheries bodies provide a mechanism for promoting country adoption of the conservation measures outlined in the voluntary FAO Code of Conduct for Responsible Fisheries (Code of Conduct) as well as the Agreement for the Implementation of the Provisions of the United Nations Convention on the Law of the Sea of 10 December 1982 relating to the Conservation and Management of Straddling Fish Stocks and Highly Migratory Fish Stocks (UN Fish Stocks Agreement). The Code of Conduct is a global instrument that aims to address the issue of ALDFG by promoting the development of environmentally safe fishing gear and practices and minimising waste. The Code of Conduct is supported by the binding UN Fish Stocks Agreement, which calls for the States to require the marking of fishing gear in accordance with international standards, a practice promoted in the prevention of ALDFG. This agreement, however, only applies to a limited number of fish species. Relevant to Africa, the General Fisheries Commission for the Mediterranean (GFCM) and South East Atlantic Fisheries Organisation (SEAFO) have adopted measures specific to the prevention of ALDFG (Gilman, 2015), but generally, Regional Fisheries Management Organisations have been slow to address the issue.

4.4.2 Sub-Regional Frameworks and Initiatives

Regional Seas Conventions Governing Africa's Marine and Coastal Areas

The UN Regional Seas Programme aims to protect and address the degradation of marine and coastal environments through engagements between groups of neighbouring coastal countries in joint, coordinated actions. There are 18 Regional Seas Programmes (RSPs), which function through action plans, and/or legally binding conventions and related protocols. Four conventions govern Africa's marine and coastal areas (Fig. 4.3):

- Abidjan Convention (Convention for Cooperation in the Protection, Management, and Development of the Marine and Coastal Environment of the Atlantic Coast of the West, Central, and Southern Africa Region);
- Nairobi Convention (Convention for the Protection of the Marine Environment and the Coastal Region of the Western Indian Ocean);
- Jeddah Convention (Regional Convention for the Conservation of the Red Sea and Gulf of Aden Environment);
- Barcelona Convention (Convention for the Protection of the Marine Environment and the Coastal Region of the Mediterranean).

These conventions form the primary initiatives relevant to marine litter prevention or removal in Africa, providing the opportunity to tailor actions to each regional context. However, they are not comprehensive in geographic coverage nor in scope of implementation. In addition, not all States are Contracting Parties (i.e. that expressed consent to be bound through, e.g. ratification) to their relevant regional convention, while others, such as Egypt, Somalia and South Africa, are Parties to more than one see Fig. 4.3 Land-locked African states[23] are not covered under these conventions due to the limited geographic scope, despite contributing to marine pollution through the transboundary transport of marine via river and other water systems (see Chap. 1, Fig. 1.1 and Chap. 2).

The scope of work of each convention varies as not all have adopted a protocol for land-based sources of pollution or a marine litter action plan. The marine litter action plans are not legally binding except for the Barcelona Convention, 2013 Regional Plan on Marine Litter Management in the Mediterranean. The fragmentation at the regional level discussed here suggests that actions and implementation will not be tackled uniformly across Africa.

Abidjan Convention

The Abidjan Convention and its associated protocol were developed in 1981 and came into force in 1984. It is a framework agreement to prevent, reduce and control marine coastal and related inland waters pollution in West, Central and Southern Africa. It has additional protocols with relevance to marine litter:

- Pointe Noire Protocol on integrated coastal zone management adopted in 2019;
- Calabar Protocol on sustainable mangrove management adopted in 2019 (the only protocol of its kind);
- Malabo Protocol on environmental norms and standards for offshore oil and gas and exploitation activities adopted in 2019;
- Grand Bassam Protocol concerning the Cooperation in the Protection and Development of the Marine and Coastal; Environment from Land-Based Sources and the Activities in the Western, Central and Southern Africa Region was signed and adopted in 2012 and
- Protocol Concerning Co-operation in Combating Pollution in Cases of Emergency in the Western and Central African Region adopted in 1985.

The Abidjan Convention hosts projects and partnerships to address marine litter, a few are mentioned here. The Abidjan Convention covered a stocktaking exercise of marine pollution in the region through three workshops organised in Namibia, Ghana and Morocco in 2019. A report of the results is currently being finalised. Another project in partnership with the African Marine Waste Network (a programme of the Sustainable Seas Trust) was an interactive webinar series hosted March–May 2021 on topics related to marine litter pollution, its impacts and actions to

---

[23] The 17 African States not covered under the four Regional Seas Conventions are Botswana, Burundi, Burkina Faso, Central African Republic, Chad, Eritrea, Eswatini, Ethiopia, Lesotho, Malawi, Mali, Niger, Rwanda, South Sudan, Uganda, Zambia and Zimbabwe.

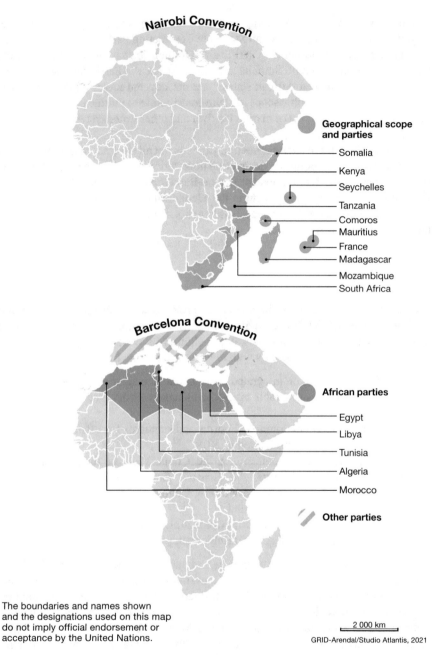

Fig. 4.3 Regional Sea Conventions and their geographical coverage

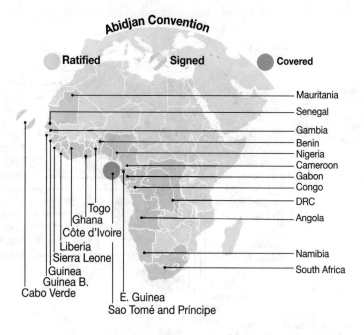

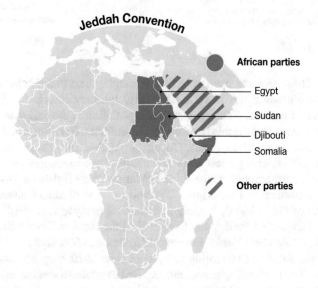

The boundaries and names shown and the designations used on this map do not imply official endorsement or acceptance by the United Nations.

2 000 km

GRID-Arendal/Studio Atlantis, 2021

Fig. 4.3 (continued)

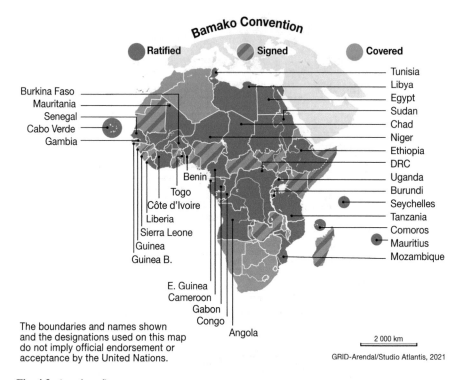

Fig. 4.3 (continued)

address it. These lectures contributed to developing best practice action plans for managing plastic waste in Africa (UNEP, 2021c). Further efforts to develop a regional action plan and national plans against plastic pollution in the region are considered under the framework of the third phase of the Multilateral Environmental Agreements in Africa, Caribbean and the Pacific Countries (ACP-MEAs) Programme funded by the EU and UNEP. This project aims to scientifically characterise plastic pollution, together with partners such as the African Marine Waste Network, WWF, GIZ and the Basel, Rotterdam and Stockholm Conventions. The Abidjan Convention further supports the call for a legally binding international treaty against plastic pollution.

The 19 Contracting Parties that have ratified the Abidjan Convention are shown in Fig. 4.3. While Cape Verde, Equatorial Guinea and São Tomé and Príncipe are located in the Abidjan Convention area, as of June 2020, they had not ratified the Convention. The contracting parties are required to establish and harmonise national laws and regulations for effective delivery of the Convention obligations.

The 12th Conference of the Parties (COP) in 2017 Decision CP.12/16: Marine waste[24] emphasises the need for the Convention Secretariat and its partners to collect and analyse data on marine waste in order to monitor future progress, carry

---

[24] http://www.abidjanconvention.org/themes/critai/documents/meetings/plenipotentiaries/wor king_documents/en/ABC-WACAF-COP12%20-Final%20Report.pdf.

out an impact assessment, inform policy at various levels and raise awareness. Data collection and analysis are encouraged to be carried out through the African Marine Waste Network using the methodology outlined in Barnardo and Ribbink (2020). While there is no specific mention of economic development and job creation, the Convention text includes recognising the economic, social and health value of the marine and coastal environments. Article 14 of the convention text highlights explicitly the need to assist in scientific and technological exchange and cooperation.

One of the challenges facing the Abidjan Convention is the lack of awareness and political will to support the Convention's activities in certain countries and limited funding (Shigwedha, 2019).

Nairobi Convention

The Nairobi Convention was adopted in 1984 and came into force in 1996, with further amendments adopted in 2010.[25] It offers a legal framework, an intergovernmental forum, and a platform for coordinated actions with partners in the West Indian Ocean region (East coast of Africa). It has additional protocols relevant to marine litter prevention and removal, including:

- the Protocol for the Protection of the Marine and Coastal Environment of the Western Indian Ocean from Land-Based Sources and Activities (Adopted March 2010); and
- the Protocol Concerning Co-operation in Combating Marine Pollution in Cases of Emergency in the Eastern African Region.

The Convention is made up of ten Contracting Parties (Fig. 4.3): Comoros, Kenya, Madagascar, Mauritius, Mozambique, Seychelles, Somalia, Tanzania, South Africa and France (La Réunion and Mayotte).

Western Indian Ocean Regional Action Plan on Marine Litter (WIO-RAPMaLi)

The Western Indian Ocean Regional Action Plan on Marine Litter (WIO-RAPMaLi)[26] was developed in 2018 in response to the UNEA resolution 1/6, 2/11 and 3/20 to address marine litter regionally, in a coordinated manner. The action plan aims to set implementation standards for contracting parties to the Nairobi Convention towards agreed commitments. There are six focus areas for actions: (i) stakeholder engagement, (ii) policy and legal frameworks, (iii) prevention and removal of marine litter, (iv) education and outreach, (v) monitoring, research and

---

[25] https://www.nairobiconvention.org/clearinghouse/sites/default/files/UNEP-DEPI-EAF-COP8-2015-10-en-Amended-Nairobi-Convention.pdf.

[26] https://nairobiconvention.org/Meeting%20Documents/December%202018/WIO-RAPMaLi_Full%20Revised%20Draft_29102018_Final.pdf.

reporting and (vi) capacity development. Interestingly, job creation and socioeconomic development are not included as focus areas of the action plan, even though they could be seen as positive reinforcement of all other focus areas.

Regional Group of Experts on Marine Litter and Microplastics

A linked success is the establishment of the regional Group of Experts on Marine Litter and Microplastics,[27] in response to "Decision CP.9/3 Management of marine litter and municipal wastewater in the Western Indian Ocean" from the ninth COP in August 2018. The Secretariat established this group in collaboration with the Western Indian Ocean Marine Science Association (WIOMSA). The main objective is to provide a knowledge exchange platform, provide policy guidance to the Nairobi Convention and synthesise research topics of relevance.

Jeddah Convention

In 1974, the Programme for the Environment of the Red Sea and Gulf of Aden (PERSGA) was initiated in collaboration with the Arab League Educational, Cultural and Scientific Organization (ALECSO) and UNEP to address transboundary threats such as marine pollution, overfishing and depletion of marine resources. In 1982, the programme was underpinned by signing the Jeddah Convention, which provides the platform for governments and partners to commit to joint and coordinated efforts to address threats to marine and coastal environments in the region.

Along with the Jeddah Convention, the "Action Plan for the Conservation of the Marine Environment and Coastal Areas in the Red Sea and Gulf of Aden", which is also legally binding, was signed in 1982. In 2005, two additional protocols were signed:

- the Protocol Concerning the Conservation of Biological Diversity and the Establishment of Network of Protected Areas in the Red Sea and Gulf of Aden; and
- the Protocol Concerning the Protection of the Marine Environment from Land-Based Activities in the Red Sea and Gulf of Aden.

Marine litter is further addressed in the Regional Marine Litter Program, which includes monitoring, capacity building and training, awareness-raising and developing guidelines and action plans for marine litter assessments. These activities are conducted both nationally and regionally.

---

[27] http://nairobiconvention.org/Meeting%20Documents/June%202019/Marine%20Litter%20R egional%20Technical%20Working%20Group%20in%20the%20Western%20Indian%20Ocean% 20region./Group%20of%20Experts%20on%20Marine%20Litter%20and%20Microplastics_Final. pdf.

The Regional Action Plan for the Sustainable Management of Marine Litter in the Red Sea and Gulf of Aden highlights seven strategies for comprehensive action to address marine litter: (i) an integrated management framework; (ii) awareness and education; (iii) legal and institutional framework; (iv) encouraging public–private partnerships; (v) removal of marine litter; (vi) research and monitoring and (vii) capacity building and training (PERSGA, 2018). Member States are expected to develop National Action Plans based on the Regional Action Plan, detailing how and when actions will be implemented. While the economic impacts of marine litter are acknowledged in the rationale and included in the awareness-raising component of the action plan; there are no specific actions to promote economic opportunities including job creation and sustainable livelihoods, which could be viewed as one of the major gaps of this action plan. Furthermore, there is no mention of financing for waste management activities and infrastructure. However, this may be indirectly promoted through the proposed public–private partnerships. Another potential gap is the lack of emphasis on science–policy interface when considering the links between the legal and institutional framework and the research and monitoring components.

Four of the seven Contracting Parties are African States (Fig. 4.3): Djibouti, Egypt, Somalia and Sudan. Other Contracting Parties are Jordan, Saudi Arabia and Yemen.

Barcelona Convention

The Barcelona Convention was adopted in 1976, entered into force in 1978, and amended in 1995 (which entered into force in 2004). The Mediterranean Action Plan (MAP) is the legally binding framework that holds the Barcelona Convention and its seven protocols. Many of these protocols are relevant to marine litter management and prevention:

- the Protocol for the Prevention of Pollution of the Mediterranean Sea by Dumping from Ships and Aircraft;
- the Protocol for the Protection of the Mediterranean Sea Against Pollution from Land-Based Sources;
- the Protocol Concerning Mediterranean Specially Protected Areas;
- the Protocol on the Prevention of Pollution of the Mediterranean Sea by Transboundary Movements of Hazardous Wastes and their Disposal and
- the Protocol on Integrated Coastal Zone Management in the Mediterranean.

The work of the MAP is guided by its 6-year mid-term strategy–the previous strategy for 2016–2021 was adopted by the 19th COP in Decision IG.22/1.[28] The strategy for the following 6 years has recently been agreed upon at the latest COP meeting in 2021. Under one of the core themes of the strategy–land and sea-based pollution–the prevention, reduction and control of marine and coastal litter and its

---

[28] UNEP/MAP Mid-Term Strategy 2016–2021. UNEP(DEPI)/MED IG.22/28. Retrieved from: https://wedocs.unep.org/bitstream/handle/20.500.11822/6071/16ig22_28_22_01_eng.pdf. Accessed: 15 June 2021.

impacts are highlighted as strategic objectives. Furthermore, under the cross-cutting themes of sustainable consumption and production (SCP), key economic sectors and lifestyles are identified as one of the upstream drivers of marine litter to be addressed. Strengthening of technical capacities of businesses, finance agents and civil society to implement SCP solutions is highlighted as another lever to prevent and reduce marine litter. Notable outputs related to marine litter include national monitoring, pilot projects to reduce upstream sources and knowledge exchange for best practices. While the financing of waste management activities and job creation are not identified explicitly as key focus areas, they are addressed indirectly though the private sector and SCP activities. The science–policy interface is specifically mentioned as a priority, together with facilitating stakeholder engagement.

Five of the 22 Contracting Parties to the Barcelona Convention are African countries (Fig. 4.3): Algeria, Egypt, Libya, Morocco and Tunisia. Other contracting parties are Albania, Bosnia and Herzegovina, Croatia, Cyprus, France, Greece, Israel, Italy, Lebanon, Malta, Monaco, Montenegro, Slovenia, Spain, Syrian Arab Republic, Turkey and the European Union.

Relevant Sub-Regional Economic Bodies, Commissions, and Frameworks

Some regional economic bodies have adopted measures to address pollution, single-use plastic production and consumption and the management and trade of plastic waste or hazardous waste. These bodies are established to engage and regulate common interests in commercial and industrial relations amongst countries. The geographic scope of some of these intergovernmental communities also extends to the non-coastal African States, which provides an opportunity to engage land-locked countries on relevant transboundary issues of marine pollution. One of them is transport through transboundary waterways–an opportunity that is not afforded in the Regional Seas Conventions.

East African Community (EAC)[29]

The Treaty for the establishment of the East African Community[30] promotes the interconnected use of national communication systems to identify polluted marine areas (Article 93 (o)); encourages joint actions to address inland water pollution monitoring and control (Article 94 (p)); encourages coordinated effort to protect the environment against all forms of pollution (Article 111 (b)). In Article 112, further supports measures to control transboundary water pollution from developmental activities (1c); encourages the adoption of common standards for control of

---

[29] Burundi, Kenya, Rwanda, South Sudan, Tanzania, and Uganda.

[30] Treaty for the establishment of the East African Community, text retrieved from: https://www.eacj.org/?page_id=33. Accessed 16 June 2021.

land and water pollution resulting from urban and industrial activities (2 h) and encourages the manufacture and use of biodegradable packaging (2c). The EAC also adopted the Polythene Materials Control Bill 2016,[31] which provides a framework to prohibit the manufacture, sale, use and importation of polythene materials on a national level in the region. The Bill is currently awaiting assent by EAC Heads of State (UNEP, 2018b).

Economic Community of West African States (ECOWAS)

The Economic Community of West African States (ECOWAS) Integrated Maritime Strategy (EIMS)[32] includes provisions to reduce, combat and control marine and coastal pollution from the maritime industry. Waste management, in particular, is addressed in the draft regional strategies on plastic waste management, e-waste and hazardous waste (UNEP, 2018b). In particular, plastic waste management issues are to be included in the revised ECOWAS Environmental Action Plan (EAP).

Southern African Development Community (SADC)

The Southern African Development Community (SADC) Regional Indicative Strategic Development Plan (RISDP) for 2020–2030 (SADC, 2020) highlights the sustainable development of the blue, green and circular economies in its third strategic objective. The key interventions to achieve this include a regional Waste Management Programme, as well as the development of a blue economy and circular economy strategies.

4.4.3   Relevant Regional and Sub-Regional Marine Litter Initiatives

In addition to these policy agreements, regional and sub-regional programmes provide environmental policy guidance and promote better practices for the marine environment, resulting in region-wide multi-partner projects that respond to specific marine pollution issues in Africa, some of which are outlined below.

---

[31] EAC Polythene Materials Control Bill, 2011. Retrieved from: https://www.eala.org/uploads/Scan_20170606_(7).pdf. Accessed: 16 June 2021.

[32] EIMS draft. 2016. Retrieved from: https://edup.ecowas.int/allevents/categories/key-resources/eims/. Accessed: 17 June 2021.

IUCN World Conservation Congress and Close the Plastic Tap Programme

The IUCN hosts the World Conservation Congress, a global agenda-setting forum on nature conservation. During its 2021 session, IUCN membership, including 45 State and government agencies from Africa, will consider and approve appropriate motions. These include three, targeted at slowing the global plastic pollution crisis in the marine environment, eliminating plastic pollution in protected areas, with priority action on single-use plastic products and avoiding unintended impacts of plastics substitution. The motions have been co-sponsored by 11 organisations from Benin, Burkina Faso, Cameroon, Côte d'Ivoire, Kenya, Malawi, Morocco and Uganda and 36 organisations from outside the continent.

The IUCN also implements the "Close the Plastic Tap" global programme, which undertakes analytical studies and supports policy and programmatic action in Africa. Current and past actions focused on Algeria, Egypt, Kenya, Libya, Morocco, and Western Sahara, Mozambique, Senegal, South Africa and Tanzania.

UNEP/IUCN National Guidance for Plastic Pollution Hotspotting and Shaping Action

Through Resolution No. 6/2019 on marine plastic litter and microplastics, the member states to the UN Environment Assembly highlighted the importance of a harmonised methodology to measure plastic flows and leakage along the value chain. This resolution laid the basis for UNEP and IUCN to develop a methodology to support countries to address existing knowledge gaps in understanding the magnitude of the challenge at a national level and thus to address the root causes of the problem. Through its Life Cycle Initiative, UNEP and the IUCN have co-developed the 'UNEP/IUCN National Guidance for Plastic Pollution Hotspotting and Shaping Action' that helps States quantify and qualify plastic pollution and offers an effective interface between science-based assessments, policy-making and action. The results from piloting the guidance in Kenya, Mozambique, Tanzania and South Africa demonstrate a clear case for African countries to address street littering, illegal dumping and open burning of waste. Addressing the lack of waste segregation combined with the separate collection, sorting and processing are quick wins and cost-effective solutions that will reduce mismanaged plastic in general waste. The financing required to implement such policy measures, including implementation, could be much lower than that needed for scaling waste management infrastructure across Africa's cities and municipalities (UNEP, 2020).

WWF's "No Plastic in Nature" Global Initiative and Regional Strategy for Africa

WWF is an independent conservation organisation working with governments, businesses, communities and civil society in more than 100 countries to sustain the natural world for the benefit of people and nature. Amongst WWF's broad programmes of work, the "No Plastic in Nature" Initiative works across the life cycle of plastics to reduce the amount of virgin plastic production, increase the circulation of material already in the system and eliminate plastic leakage. The initiative focuses on three core pillars: global governance, business engagement and Plastic Smart Cities. A regional 2020–2025 strategy for WWF offices in Africa has been developed to implement the initiative across the region. Capacity development is one of the key focuses alongside evidence-based campaigns to encourage governments to improve enforcement of existing legislation and implement more robust policies, including but not limited to extended producer responsibility (EPR). Through its policy advocacy, WWF encourages governments to support the mandate to start negotiations on a new global legally binding treaty to address plastic pollution.

Sustainable Seas Trust and the African Marine Waste Network

The Sustainable Seas Trust is a nonprofit organisation that focuses on research, education, awareness-raising and enterprise development across Africa. One of its core programmes is the African Marine Waste Network (AMWN),[33] a collaborative platform for knowledge exchange within Africa and beyond. It aims to mobilise resources to enable stakeholders to reduce marine pollution through research, education, capacity building, enterprise development and communications. The AMWN also partners with WIOMSA to monitor marine litter throughout the WIOMSA region (see Sect. 4.4.3.5).

Western Indian Ocean Marine Science Association (WIOMSA)

The WIOMSA[34] is a nonprofit membership organisation established in 1993 that covers the same geographic scope as the Nairobi Convention. Its activities include promoting education, science and technological development in marine sciences and particularly the interface between research, management and governance of marine and coastal ecosystems. As part of its Marine and Coastal Science for Management programme, WIOMSA, in collaboration with the Sustainable Seas Trust through the AMWN and country partners, is monitoring marine litter in Kenya, Madagascar, Mauritius, Mozambique, Seychelles, Tanzania and South Africa. Baselines will be determined at each site for the relevant targets set in SDG 14.1. A marine litter

---

[33] https://sst.org.za/projects/african-marine-waste-network/.

[34] https://www.wiomsa.org/about-wiomsa/.

monitoring manual has been developed describing guidelines to measure litter in rivers, estuaries, beaches and on land (Barnardo & Ribbink, 2020).

High-level panel for a sustainable ocean economy in Africa

As part of the global High-Level Panel (HLP) on a Sustainable Ocean Economy, the three Sherpas[35] representing the African continent–Ghana, Namibia, and Kenya–coordinate and host meetings for their respective regional blocs, which are eventually intended to culminate in an Africa-wide position to be discussed and adopted at an AU Heads of State Summit meeting. A Western Indian Ocean regional Meeting was held in Mombasa in December 2019, which provided an interactive knowledge-sharing platform forged a shared understanding of ocean-related issues critical to Africa and developed A Call to Action to Save African Fisheries.

The HLP has noted the issue of pollution, especially from the tonnes of plastics that find their way into the ocean each year and the threats it poses to ecosystems, health and livelihoods. It has included pollution and plastic waste as one of three promising pathways to address Africa's ocean-related challenges.

Ellen MacArthur Foundation Plastics Pact Network in Africa

The Ellen MacArthur Foundation's Plastics Pact Network[36] is a coordinated response to plastic pollution, specifically plastic packaging, via national or regional initiatives that bring together stakeholders to implement context-specific actions towards a circular economy for plastics. Each Plastic Pact is led by a local organisation bringing together actors from across the entire plastics packaging value chain, including government bodies, businesses and civil society, behind a shared vision with an ambitious set of targets. The targets are specific to each context but all align towards eliminating unnecessary and problematic plastic packaging, ensuring plastic packaging is reusable, recyclable or compostable and increasing effective recycling rates and use of recycled content in plastic packaging.

The South African Plastics Pact was the first of its kind in Africa, launched by WWF in partnership with the South African Plastics Recycling Organisation and WRAP in January 2020 and now implemented by GreenCape. Two other Plastics Pacts in Africa are currently developing in Kenya and Senegal.

Abandoned, lost, or otherwise discarded fishing gear (ALDFG) initiatives in Africa

The Secretariat of the Basel, Rotterdam and Stockholm Conventions in cooperation with S-cycles, the Öko-Institut, Grid-Arendal and the Ministry of Environment, Science, Technology and Innovation of Ghana, with funding from the Norwegian Ministry of Foreign Affairs, implemented a pilot project to collect and recycle waste fishing nets together with local fishers in Accra, Ghana. A Global Ghost Gear Initiative (GGGI) project, under the SOFER initiative, has a programme called the Fishing Net Gains Nigeria, which aims to create economic opportunities for coastal

---

[35] Sherpas are advisors to the Members of the High-Level Panel for a Sustainable Ocean Economy.

[36] https://www.ellenmacarthurfoundation.org/our-work/activities/new-plastics-economy/plastics-pact.

communities, including women and youth, through developing best practices for addressing ghost gear in the region. The programme includes training women to design, make and sell crafts from waste fishing gear as a source of income. The craft workshops will be supplied with materials collected at fishing gear collection sites. Volunteer divers will be trained to collect data and retrieve ghost gear where safe and feasible.

Box 4.1: Bans on Plastics Carrier Bags and Other Single-Use Plastic Items

The banning or regulation of single-use plastic items, especially plastic carrier bags, has been a popular legislative instrument to reduce the generation of plastic waste and pollution. Reasons include malaria outbreaks due to stagnant water collected in littered plastic bags in Kenya, sewage systems clogged by plastic bags causing floods in Cameroon and death of livestock due to ingestion of plastic bags in Mauritania (Larsen & Venkova, 2017).

Over 30 African States have some legislation to regulate or completely ban the use, manufacture, sale, free distribution and import of plastic carrier bags (UNEP, 2018a). Examples include:

- The EAC Polythene Materials Control Bill 2016 provides a legal framework for prohibiting of the manufacture, sale, import and use of polythene materials, including plastic carrier bags, intended for national-level implementation for countries based in the EAC region.[37]
- Rwanda was one of the first and most strict amongst the African states, with its ban enforced in 2008, including confiscation of plastic bags upon arrivals at airports.
- Kenya's ban on single-use plastic bags in 2017 was followed by a ban in June 2020 on other single-use plastic items in protected natural areas such as National Parks, beaches, forests, and conservation areas.
- Countries such as South Africa and Cameroon have adopted taxes on certain types of bags as a disincentive instead of an outright ban, transferring some of the environmental cost onto the consumer.
- In 2021, South Africa amended its plastic carrier bag regulations[38] by setting a minimum recycled content requirement to be enforced in a phased approach. This aims to promote end-use markets to recycled content, which has been a major barrier for the country's recycling economy (Van Os & De Kock, 2021).
- In 2019, in collaboration with the AU First Ladies, the AU Commission hosted a high-level working session on the "Banning of plastics in Africa towards a pollution-free Africa".
- In Côte d'Ivoire, the plastics industry actors opposed the proposed ban due to the threat of job losses and other economic repercussions. This caused several delays in enforcement, as well as the addition of exemptions for

biodegradable plastic bags and specific applications (Kobo, 2014; Excell, et al. 2018).

- In Senegal, the plastic ban in 2015 had failed to be implemented; however, the state proposed another plastic ban in 2020. The proposed ban included water sachets and was protested by various actors selling drinking water in plastic sachets. It was estimated that the ban would potentially put 30,000 jobs at risk (Oladipo & Niang, 2021).

- In 2012 (with an effective date of 1 January 2013 and a 6-month moratorium) the Mauritanian government decided to ban the manufacture, marketing and use of plastic bags in order to protect the environment. But due to the lack of accompanying measures (alternatives, lack of awareness and public acceptance, monitoring, sanctions…) and with the pressure of certain lobbies (traders, importers, bag producers…) the measure has remained ineffective.

The successful implementation of plastic bag bans varies across the continent and has been the main criticism of its effectiveness. Several reports of enforcement issues of the plastic bag bans are due to illegal trade across "porous" national borders as well as established informal markets (UNEP, 2018c). Some of the potential unintended consequences of bans, such as the loss of jobs and revenue in the plastic bag production industry, are also a concern (Godfrey, 2019), especially for African countries with a plastic production value chain such as Nigeria and South Africa.

Besides socioeconomic impacts, life-cycle sustainability impacts should be considered, primarily to assess the risk of introducing alternative products and materials into the economy to offset the ban. A life-cycle assessment study on single-use plastic carrier bags and their alternatives in South Africa revealed some of the cautions necessary when replacing them with alternative materials such as paper, biodegradables or reusable polyester bags (Russo et al., 2020). This study further affirmed the need use recycled content to replace virgin plastic as more favourable in life cycle sustainability impacts.

The banning of plastic goods may be seen as a knee-jerk reaction to the broader issue of plastic pollution and marine litter, as it is limited to a singular focus on certain types of plastic waste. A systems approach to policy and legislative instruments to address plastic pollution and marine litter is necessary to avoid fragmented interventions and unintended consequences.

---

[37] Burundi, Kenya, Rwanda, South Sudan, Tanzania, and Uganda.

[38] South Africa amendments to the plastic carrier bag regulations. 2021. https://www.environment. gov.za/sites/default/files/gazetted_notices/nema_amendements_44421gon317.pdf.

4.5   Implementation Challenges

The international and regional legal and policy frameworks assessment shows that the duty to prevent marine litter from land- and sea-based sources has been clearly established. Less clear, however, is the duty to provide sustainable funding for such purposes while ensuring a safe and healthy environment and access to sustainable livelihoods. The SDGs provide the most robust guidance in this regard, providing a holistic approach to preventing marine litter through environmentally sustainable waste management that supports the principles of justice for all. The drivers of marine litter in Africa are complex, supporting the need to incorporate a broader range of measures than those included in global and regional instruments for preventing pollution, managing chemicals and waste and protecting species and biodiversity.

By recognising the common challenges and limitations of implementation at the national level, research and sharing best practices can facilitate the cost-effective transformation of the plastics value chain in Africa, with a primary focus on adding value to waste and providing sustainable financing of waste management systems. Examples of success exist on the continent that can stimulate action in neighbouring countries. Still, experiences have also shown that a coordinated and harmonised regional approach is required to support compliance and enforcement. By incorporating all 17 SDGs into the systemic transformation of waste generation and management, the co-benefits of poverty reduction and environmental justice can be realised within the African context.

National policy and regulatory frameworks could prioritise stringent provisions that aim to address street littering, open and illegal dumping, open burning of waste and waste segregation combined with the separate collection, sorting and processing as a quick win, cost-effective solutions that contribute to minimising the mismanaged plastic component in general waste. The financing required to implement such policy measures targeted at source reduction, including implementation, could be much lower than that required for scaling waste management infrastructure across Africa's cities and municipalities. Lebreton and Andrady (2019) suggest that the gradual increase in waste management infrastructure may not be enough for some parts of the world, particularly in Africa, by 2060. The investment in recycling infrastructure appears uneconomic because of the substantially higher recycling cost than landfilling (Goddard, 1995).

There is increased appreciation of the circular economy concept across Africa, which is being recognised in policy and legislative formulations. Even though the concept presents economic and social opportunities, it remains weak in addressing the social equity gap. The number of citizen-driven initiatives in Africa engaged in beach clean-ups or collection and trade in waste is on the rise. These present opportunities for addressing the social equity gap. States though need to explore appropriate integration models that ensure that these informal sector actors can be integrated into the formal waste economy to safeguard their interests and ensure that the benefits from the waste economy impact the broadest possible segment of society. From a waste prevention perspective, African countries need to develop and support the implementation of policies that emphasise the creation and marketing of recyclable or recoverable materials, either at the same or higher quality, coupled with the sourcing of renewable and secondary material in the manufacture of plastic objects. There is also a need to explore other innovative approaches to addressing plastic pollution by incentivising the reuse of plastics and promoting other measures that address plastic waste beyond recycling, which tends to be the key strategy at the heart of the circular economy. Such policy reforms need to support the establishment of multi-level, multi-sector platforms at the national level to ensure unified strategies across the entire life-cycle stages of plastics as no single actor can independently drive complete life-cycle improvements.

4.6   Recommended Best Practices for Prevention and Reduction of Marine Litter in Africa

A number of response options are available to governments operating at the national and sub-national levels. These must be suited to the local context and supported by socioeconomic studies to ensure the intended outcomes are met and various industries and communities are not negatively affected. Important to the design of policies to prevent marine litter is the inclusion of the producer in the physical and financial responsibility for the waste resulting from their products placed on the market. The cost of waste management can be borne by a mix of local government, producers and consumers. Examples of consumer contribution to the reduction of marine litter through improved waste management include pay-as-you-throw schemes to fund household waste collection, as well as shopping bag taxes to reduce their use and fund various waste management services.

> **Box 4.2: EPR in the African Context—The Example of South Africa (Climate Legal, 2020)**
> EPR schemes can assist in preventing marine litter in the short term by improving formal collection systems and in the long term by incentivising design for the environment. A central aim of EPR schemes is to reduce the burden on local councils of end-of-life treatment of products and transfer the costs of these services to producers and/or consumers. In addition, economic opportunities and environmental benefits can be gained (OECD, 2016). EPR schemes can be voluntary, co-regulatory or mandatory and include product take-back schemes (e.g. producer is responsible for collection and recycling), market-based instruments (e.g. deposit-refund, advanced disposal fees and taxes), performance standards (e.g. minimum recycled content) and information (e.g. labelling requirements) (OECD, 2016). Schemes should determine if producers regulated by the scheme will cover the full cost or partial costs.

In South Africa, an industry-led self-regulated initiative for PET recycling, PETCO, has been in operation since 2004. No mandatory EPR schemes had been adopted in Africa by 2013 (OECD, 2016). However, this situation has changed in recent years with the adoption of legislation that provides for the development of such schemes. Examples include South Africa, where EPR regulations have recently been adopted, and Kenya (see Opondo, 2020), where a draft bill is still under discussion.

The OECD has outlined the following recommended elements for effective EPR legislation (OECD, 2016):

Design and Governance—Provision should be made for updating targets, fees and other aspects affected by changes in the market. Enforcement measures can include mandatory scheme registers, accreditation schemes, penalties for non-compliance, robust monitoring systems and reporting that is independently audited.

Definitions, Roles and Responsibilities—Stakeholders of EPR schemes include producers, national governments, local councils, retailers and consumers. Relevant stakeholders should be defined in legislation, particularly those classified as producers. The roles and responsibilities of each will vary with the product range in scope of the scheme and the objectives to be achieved by each sector and by the scheme overall. Consideration must also be given to "free-riders" (producers that are not required to contribute to the scheme but benefit from it), "orphaned" products (the producer of the product is no longer in business) and online sales from international providers that may be difficult to capture in national legislation.

Transparency and Preventing Anti-Competitive Behaviour—Schemes must be held accountable for their performance, particularly where administration is given to a Producer Responsibility Organisation (PRO). Information that can strengthen transparency includes disclosing costs and reporting on the number of products placed on the domestic market, fees charged to producers and performance regarding rates for collection, reuse, recycling, recovery and landfill. Open and non-discriminatory tenders should promote competition in the services market and allow new technologies and processes entry.

Incentivise Design for The Environment—Eco-modular fees for EPR-regulated products can promote improved design, particularly if these fees reflect the complexity and cost of end-of-life management of products. Where this relationship is not incorporated into EPR schemes, eco-design improvements have proved difficult to achieve if collective producer responsibility is applied.

Informal Sector Integration—Formal waste management systems must integrate informal waste collectors. Excluding this sector from access to waste streams should be avoided where these streams would be managed under new EPR schemes.

Stakeholder Consultation—As with introducing any new systems, whether voluntary or legislated, stakeholder engagement is essential at all phases of development and review.

A key component of the transformation of waste management in South Africa is integrating of the informal waste sector. An estimated 60,000 people work in this sector, reportedly saving local councils an estimated ZAR 700 million annually (Department of Environmental Affairs of the Republic of South Africa, 2019). The Waste Management Act of South Africa (Act No. 59 of 2008) provides for the development of mandatory EPR schemes for relevant products. However, only voluntary schemes have been implemented. The PET Recycling Company (PETCO) of South Africa was established in 2004 and provided a voluntary, industry-driven and financed EPR scheme for PET bottles. The fee paid by members is voluntary and achieved a 62% collection rate in 2019 (PETCO, 2019a). PETCO's How-To Guide to Section 18 for Producers (PETCO, 2021) assists their members in understanding their obligations under the new EPR Regulations, while the guidelines on Designing for the Environment (PETCO, 2019b) promote improved design to support end-of-life processes for PET bottles. In lieu of mandatory EPR regulations in Kenya, PETCO was also established in Kenya in 2018 (PETCO Kenya, 2021) and operated in a similarly to PETCO in South Africa before the new EPR Regulations were introduced. As of 2019, five drop-off sites had been set up in Nairobi in partnership with retailers.

The recently adopted Regulations Regarding EPR in GN 1184, Government Gazette No. 43879 of 5 November 2020 National Environmental Management: Waste Act (59/2008) (EPR Regulations) in South Africa give effect to Section 18 of the Waste Management Act and are likely to move South Africa towards mandatory schemes. These regulations provide definitions and requirements of producers such as fee collection, record keeping and auditing. Producers are also required to conduct life-cycle assessments (including minimising waste and toxicity) within 5 years of establishing an EPR scheme, use environmental labels, integrate the informal sector, develop secondary markets for recyclates and work towards equal representation within the entire value chain. The EPR Regulations mandate implementation of the waste hierarchy as well as cleaner production principles. A producer registration process is provided to prevent "free-riders."

Waste pickers are well represented in the 2020 National Waste Management Strategy (DEFF, 2020a) and the 2020 Waste Picker Integration

Guideline, developed through six participatory workshops that included industry (DEFF, 2020b). The Waste Picker Integration Guideline includes a section on integration within EPR schemes. It will guide future development of such schemes, together with stakeholder engagement, particularly the South African Waste Pickers Association. The EPR legal and policy framework developed in South Africa provides a good foundation for other African countries to build on but is relatively new in South Africa. They will require investment in the infrastructure development to support the mandatory measures adopted and enforcement if they are to establish a compelling example for other countries to follow.

4.7 Integrated Waste Management Policy and Strategy

An integrated waste management strategy should include various approaches that collectively work towards the goals and objectives of the strategy. These can be binding or voluntary, and a comprehensive strategy would likely include a combination of both. The waste hierarchy should provide the overall framework for any waste management strategy as a priority. The waste hierarchy promotes the most environmentally friendly options, with the least-preferred disposal options being landfilling or dumping (US EPA, 2020). According to the waste hierarchy, reduction strategies should be prioritised to minimise waste generation at source. This includes education and behaviour change on an individual and industrial level to emphasise reducing the use of plastic products. Thereafter, reuse can extend the life of products, while recycling can reduce the need to extract raw materials. Where products are not designed for reuse or recycling, or where adequate environmentally sound waste management options are not available, trade of waste in compliance with the international regulations can provide an alternative to incineration and landfill.

4.8 Preventing Marine Litter Through a Holistic Lens

Many African States prioritise the Blue Economy principles to stimulate sustainable and integrated maritime and freshwater industries (see Chap. 1). Marine litter presents a risk to maritime industries, resulting in direct costs as well as indirect impacts, for which the costs are more difficult to calculate (McIlgorm et al., 2020). Although the Blue Economy approach shares some principles with the circular economy, the latter provides greater scope for improving livelihoods across a broader range of sectors. However, in practice, both tend to focus on economic activity, giving less emphasis to social and environmental outcomes. Many of the jobs that need to be stimulated through the promotion of circular systems may fall outside of the industries that make up the Blue Economy. It is therefore imperative that Africa seeks to broaden the focus of marine litter prevention from a purely waste management perspective towards the circular economy approach, with emphasis on resource efficiency as well as social equity aspects, such as the distribution of wealth generated through the new opportunities provided by the circular economy. By addressing the issue of marine litter through the lens of job creation, the co-benefits of improved livelihoods and the right to a safe and productive environment can be collectively targeted. Therefore, the design of legal and policy frameworks should consider the most cost-effective ways to deliver on multiple benefits, including reducing the cost of waste management to the public sector. In the short term, the consumer can play a vital role in sorting and disposing of wastes appropriately, participating in clean-ups, spreading awareness and promoting ownership of the environment.

Acknowledgements We would like to acknowledge the valuable insights proved by Peter Ryan, Anham Salyani, Salieu Sankoh, Tony Ribbink, Abdoulaye Diagana, Alison Amoussou, as well as Jost Dittkirst and the Plastic Task Force of the Secretariat of the Basel, Rotterdam and Stockholm Conventions in peer-reviewing this chapter. We would also like to acknowledge Nieves López and Federico Labanti (Studio Atlantis) for creating the illustrations.

Annex 4.1: SDG Targets and Indicators Relevant to the Prevention of Marine Litter, Livelihoods and a Safe Environment

Target	Indicator
SDG 6. *Ensure availability and sustainable management of water and sanitation for all*	
Target 6.3—By 2030, improve water quality by reducing pollution, eliminating dumping and minimizing release of hazardous chemicals and materials, halving the proportion of untreated wastewater and substantially increasing recycling and safe reuse globally	6.3.1 Proportion of domestic and industrial wastewater flows safely treated 6.3.2 Proportion of bodies of water with good ambient water quality
SDG 8. *Promote sustained, inclusive and sustainable economic growth, full and productive employment and decent work for all*	
Target 8.3—Promote development-oriented policies that support productive activities, decent job creation, entrepreneurship, creativity and innovation and encourage the formalization and growth of micro-, small- and medium-sized enterprises, including through access to financial services	8.3.1 Proportion of informal employment in total employment, by sector and sex
Target 8.4—Improve progressively, through 2030, global resource efficiency in consumption and production and endeavour to decouple economic growth from environmental degradation, in accordance with the 10Year Framework of Programmes on Sustainable Consumption and Production, with developed countries taking the lead	8.4.1 Material footprint, material footprint per capita, and material footprint per GDP 8.4.2 Domestic material consumption, domestic material consumption per capita and domestic material consumption per GDP
Target 8.9—By 2030, devise and implement policies to promote sustainable tourism that creates jobs and promotes local culture and products	8.9.1 Tourism direct GDP as a proportion of total GDP and in growth rate
SDG 9. *Build resilient infrastructure, promote inclusive and sustainable industrialization and foster innovation*	
Target 9.5—Enhance scientific research, upgrade the technological capabilities of industrial sectors in all countries, in particular developing countries, including, by 2030, encouraging innovation and substantially increasing the number of research and development workers per 1 million people and public and private research and development spending	9.5.1 Research and development expenditure as a proportion of GDP 9.5.2 Researchers (in full-time equivalent) per million inhabitants
SDG 11. *Make cities and human settlements inclusive, safe, resilient and sustainable*	

(continued)

(continued)

Target	Indicator
Target 11.1—By 2030, ensure access for all to adequate, safe and affordable housing and basic services and upgrade slums	11.1.1 Proportion of urban population living in slums, informal settlements or inadequate housing
Target 11.6—By 2030, reduce the adverse per capita environmental impact of cities, including by paying special attention to air quality and municipal and other waste management	11.6.1 Proportion of municipal solid waste collected and managed in controlled facilities out of total municipal waste generated, by cities
SDG 12. *Ensure sustainable consumption and production patterns*	
Target 12.1—Implement the 10-Year Framework of Programmes on Sustainable Consumption and Production Patterns, all countries taking action, with developed countries taking the lead, taking into account the development and capabilities of developing countries	12.1.1 Number of countries developing, adopting or implementing policy instruments aimed at supporting the shift to sustainable consumption and production
Target 12.4—By 2020, achieve the environmentally sound management of chemicals and all wastes throughout their life cycle, in accordance with agreed international frameworks, and significantly reduce their release to air, water and soil in order to minimize their adverse impacts on human health and the environment	12.4.1 Number of parties to international multilateral environmental agreements on hazardous waste, and other chemicals that meet their commitments and obligations in transmitting information as required by each relevant agreement 12.4.2 (*a*) Hazardous waste generated per capita; and (*b*) proportion of hazardous waste treated, by type of treatment
Target 12.5—By 2030, substantially reduce waste generation through prevention, reduction, recycling and reuse	12.5.1 National recycling rate, tonnes of material recycled
Target 12.7—Promote public procurement practices that are sustainable, in accordance with national policies and priorities	12.7.1 Degree of sustainable public procurement policies and action plan implementation
Target 12.8—By 2030, ensure that people everywhere have the relevant information and awareness for sustainable development and lifestyles in harmony with nature	12.8.1 Extent to which (i) global citizenship education and (ii) education for sustainable development are mainstreamed in (*a*) national education policies; (*b*) curricula; (*c*) teacher education and (*d*) student assessment
SDG 14. *Conserve and sustainably use the oceans, seas and marine resources*	
Target 14.1—By 2025, prevent and significantly reduce marine pollution of all kinds, in particular from land-based activities, including marine debris and nutrient pollution	14.1.1 (*a*) Index of coastal eutrophication; and (*b*) plastic debris density
Target 14.2—By 2020, sustainably manage and protect marine and coastal ecosystems to avoid significant adverse impacts, including by strengthening their resilience, and take action for their restoration in order to achieve healthy and productive oceans	14.2.1 Number of countries using ecosystem-based approaches to managing marine areas

(continued)

(continued)

Target	Indicator
Target 14.C—Enhance the conservation and sustainable use of oceans and their resources by implementing international law as reflected in the United Nations Convention on the Law of the Sea, which provides the legal framework for the conservation and sustainable use of oceans and their resources, as recalled in paragraph 158 of "The future we want"	14.c.1 Number of countries making progress in ratifying, accepting and implementing through legal, policy and institutional frameworks, ocean-related instruments that implement international law, as reflected in the United Nations Convention on the Law of the Sea, for the conservation and sustainable use of the oceans and their resources

Annex 4.2 List of International and Regional Policy Instruments, Agreements and Declarations Relevant to Marine Plastic Litter

1972 Convention on the Prevention of Marine Pollution by Dumping of Wastes and Other Matter, opened for signature.
13 November 1972, 1046 UNTS 120 (Entered into Force 30 August 1975) ('London Convention') <https://treaties.un.org/%20doc/Publication/UNTS/Volume%201046/volume-1046-I-15749-%20English.pdf Doc/Publication/UNTS/Volume1046/volume-1046-I-15749-English.Pdf>.
1978 Protocol of 1978 relating to the International Convention for the Prevention of Pollution from Ships of 2 November 1973, as amended, opened for signature 17 February 1978, 1340 UNTS 184 (entered into force 2 October 1983) ('MARPOL 73/78') <https://treaties.un.org/doc/Publication/UNTS/Volume%201340/volume-1340-I-22484-English.pdf>.
1979 The Convention on the Conservation of Migratory Species of Wild Animals opened for signature 23 June 1979, [1991] ATS 32 (entered into force 1 November 1983) ('CMS') <http://www.austlii.edu.au/au/other/dfat/treaties/1991/32.html>.
1982 United Nations Convention on the Law of the Sea opened for signature 10 December 1982, 1833 UNTS 3 (entered into force 16 November 1994) ('Law of the Sea Convention') <https://www.un.org/depts/los/convention_agreements/texts/unclos/unclos_e.pdf> .
1989 Basel Convention on the Control of Transboundary Movements of Hazardous Wastes and Their Disposal, opened for signature 22 March 1989, 1673 UNTS 57 (entered into force 5 May 1992) ('Basel Convention') <https://www.basel.int/Portals/4/BaselConvention/docs/text/BaselConventionText-e.pdf> .
1991 Convention on Environmental Impact Assessment in a Transboundary Context (ECE/MP.EIA/21), opened for signature 25 February 1991, 1989 UNTS 309 (No. 34028) (entered into force 10 September 1997) ('Espoo Convention') <http://www.unece.org/index.php?id=40450&L=0> .
1992 Convention on Biological Diversity, opened for signature 5 June 1992, 1760 UNTS 79 (entered into force 29 December 1993) ('Convention on Biological Diversity') <https://www.cbd.int/convention/text/default.shtml> .
1995 The Agreement for the Implementation of the Provisions of the United Nations Convention on the Law of the Sea of 10 December 1982 relating to the Conservation and Management of Straddling Fish Stocks and Highly Migratory Fish Stocks, opened for signature 4 December 1995, 2167 UNTS 3 (entered into force 11 November 2001) ('United Nations Fish Stocks

Agreement') <https://treaties.un.org/doc/Treaties/1995/08/19950804 08–25 AM/Ch_XXI_07p.pdf> .

1996 Protocol to the Convention on the Prevention of Marine Pollution by Dumping of Wastes and Other Matter, 1972, opened for signature 7 November 1996, 36 ILM 1 (1997) (entered into force 24 March 2006) ('London Protocol') <http://www.austlii.edu.au/au/other/dfat/treaties/2006/11.html> .

1997 Convention on the Law of the Non-Navigational Uses of International Watercourses, opened for signature 21 May 1997, UN Doc A/RES/51/229 (entered into force 17 August 2014) ('UN Watercourse Convention') <http://www.un.org/documents/ga/res/51/ares51-229.htm> .

2001 Stockholm Convention on Persistent Organic Pollutants, opened for signature 22 May 2001, 2256 UNTS 119.

(entered into force 17 May 2004) ('Stockholm Convention') <https://treaties.un.org/doc/Treaties/2001/05/2001052212–55 PM/Ch_XXVII_15p.pdf> .

2011 Regulations for the Prevention of Pollution by Garbage from Ships (Resolution MEPC.201(62)), opened for signature 15 July 2011, (entered into force 1 January 2013) ('MARPOL Annex V') <http://www.imo.org/en/OurWork/Environment/PollutionPrevention/Garbage/Documents/2014 revision/ RESOLUTION MEPC.201(62) Revised MARPOL Annex V.pdf > .

2012 CBD, Marine and coastal biodiversity: sustainable fisheries and addressing adverse impacts of human activities, voluntary guidelines for environmental assessment, and marine spatial planning, UNEP/CBD/COP/DEC/XI/18, 11, (CBD Decision XI/18) <https://www.cbd.int/doc/decisions/cop-11/cop-11-dec-18-en.pdf>.

2014 CBD, Marine and coastal biodiversity: Impacts on marine and coastal biodiversity of anthropogenic underwater noise and ocean acidification, priority actions to achieve Aichi Biodiversity Target 10 for coral reefs and closely associated ecosystems, and marine spatial planning and training initiatives, UNEP/CBD/ COP/DEC/XII/23, 12 (Marine and coastal biodiversity: Impacts on marine and coastal biodiversity of anthropogenic underwater noise and ocean acidification, priority actions to achieve Aichi Biodiversity Target 10 for coral reefs and closely associated ecosystems, and marine spatial planning and training initiatives) < https://www.cbd.int/doc/decisions/cop-12/cop-12-dec-23-en.pdf > .

2016 CBD, Addressing impacts of marine debris and anthropogenic underwater noise on marine and coastal biodiversity, CBD/COP/DEC/XIII/10, 13, (CBD Decision XIII/10) <https://www.cbd.int/doc/decisions/cop-13/cop-13-dec-10-en.pdf> .

2017 United Nations General Assembly (UNGA), Sustainable fisheries, including through the 1995 Agreement for the Implementation of the Provisions of the United Nations Convention on the Law of the Sea of 10 December 1982 relating to the Conservation and Management of Straddling Fish Stocks and Highly Migratory Fish Stocks, and related instruments, A/RES/71/123, 71, (UNGA Resolution 71/123) <http://undocs.org/A/RES/71/123> .

1980 Protocol for the Protection of the Mediterranean Sea against Pollution from Land-Based Sources and Activities, as amended 7 March 1996, opened for signature 7 March 1996, 1328 UNTS 120 (entered into force 11 May 2008) ('LBS/A Protocol for the Mediterranean') <http://wedocs.unep.org/bitstream/handle/20.500.11822/7096/Consolidated_LBS96_ENG.pdf?sequence%20=%205&isAllowed%20=%20y> .

1981 Convention for Co-operation in the Protection and Development of the Marine and Coastal Environment of the West and Central African Region, opened for signature 23 March 1981, 20 ILM (1981) 746 (entered into force 05 August 1984) ('Abidjan Convention') <http://abidjanconvention.org/index.php?option=com_content&view=article&id=100&Itemid=200&lan=en> .

1991 Bamako Convention on the Ban of the Import into Africa and the Control of Transboundary Movement and Management of Hazardous Wastes within Africa, opened for signature 30 January 1991, 2101 UNTS 211 (entered into force 22 April 1998) ('Bamako Convention')

<https://www.opcw.org/chemical-weapons-convention/related-international-agreements/toxic-chemicals-and-the-environment/bamako-convention/> .

2010 Protocol for the Protection of the Marine and Coastal Environment of the Western Indian Ocean from Land-Based Sources and Activities, opened for signature 31 March 2010, ('LBS/A Protocol for the Western Indian Ocean') <http://www.unep.org/nairobiconvention/protocol-protection-marine-and-%20coastal-environment-wio-land-based-sources-and-activities> .

2012 Additional Protocol to the Abidjan Convention Concerning Cooperation in the Protection and Development of Marine and Coastal Environment from Land-Based Sources and Activities in the Western, Central and Southern African Region (UNEP(DEPI)/WACAF/LBSA/MOP1/2), opened for signature 22 June 2012, ('LBS/A Protocol of Western, Central and Southern African Region') <http://abidjanconvention.org/media/documents/protocols/LBSA Protocol-Adopted.pdf> .

2013 Regional Plan on Marine Litter Management in the Mediterranean in the Framework of Article 15 of the Land Based Sources Protocol (Decision IG.21/7), opened for signature 6 December 2013, (entered into force 8 July 2014) ('Action Plan for Marine Litter in the Mediterranean') <http://www.unepmap.org/index.php?module%20=%20content2&catid%20=%20001,011,006> .

2005 United Nations General Assembly, Sustainable fisheries, including through the 1995 Agreement for the Implementation of the Provisions of the United Nations Convention on the Law of the Sea of 10 December 1982 relating to the Conservation and Management of Straddling Fish Stocks and Highly Migratory Fish Stocks, and related instruments, A/RES/60/31, (UNGA Resolution 60/31) (29 November 2005) <http://www.un.org/depts/los/general_assembly/general_assembly_resolutions.htm> .

2012 Manila Declaration, Manila Declaration on Furthering the Implementation of the Global Programme of Action for the Protection of the Marine Environment from Land-based Activities, UNEP/GPA/IGR.3/CRP.1/Rev.1, (Manila Declaration) (27 January 2012) <http://www.unep.org/regionalseas/globalmeetings/15/ManillaDeclarationnew.pdf> .

2011 The Honolulu Strategy, A Global Framework for Prevention and Management of Marine Debris, 25 March 2011, (Honolulu Strategy) <http://www.unep.org/gpa/documents/publications/honolulustrategy.pdf> .

1995 GPA, Global Programme of Action for the Protection of the Marine Environment from Land-based Activities (GPA), UNEP(OCA)/LBA/IG.2/7, (GPA) (3 November 1995)

1985 Montreal Guidelines for the Protection of the Marine Environment against Pollution from Land-Based Sources, Decision 13/18/II, (Montreal Guidelines for LBS) (24 May 1985) <http://www.unep.org/law/PDF/UNEPEnv-%20LawGuide&PrincN07.pdf>.

1995 FAO Code of Conduct for Responsible Fisheries, ('Code of Conduct'). <http://www.fao.org/docrep/005/v9878e/v9878e00.HTM>.

2015 UNGA, Transforming our world: the 2030 Agenda for Sustainable Development, A/Res/70/1, (The 2030 Agenda) <https://undocs.org/A/RES/70/1>.

2002 United Nations, Johannesburg Declaration on Sustainable Development (A/CONF.199/20) Chap. 1, Resolution 1, (Johannesburg Declaration on Sustainable Development (A/CONF.199/20) Chap. 1, Resolution 1) <https://documents-dds-ny.un.org/doc/UNDOC/GEN/N02/636/93/PDF/N0263693.pdf?OpenElement>.

References

Africa Business. (2021). *The plastics industry in Africa*. https://www.africa-business.com/features/plastics.html

AU. (2012). *2050 Africa's integrated maritime strategy* (2050 AIM strategy). https://wedocs.unep.org/bitstream/handle/20.500.11822/11151/2050_aims_srategy.pdf

AU. (2015). *First ten year implementation plan*. Agenda 2063. https://au.int/sites/default/files/documents/33126-doc-11_an_overview_of_agenda.pdf

AU-IBAR. (2019). *Africa blue economy strategy*. Nairobi, Kenya. https://www.infoafrica.it/wp-content/uploads/2020/07/sd_20200313_africa_blue_economy_strategy_en.pdf

Babayemi, J. O., Nnorom, I. C., Osibanjo, O., & Weber, R. (2019). Ensuring sustainability in plastics use in Africa: Consumption, waste generation, and projections. *Environmental Sciences Europe, 31*, 60. https://doi.org/10.1186/s12302-019-0254-5

Barnardo, T., & Ribbink A. J. (2020). *African marine litter monitoring manual*. African Marine Waste Network, Sustainable Seas Trust, Port Elizabeth. https://www.wiomsa.org/wp-content/uploads/2020/07/African-Marine-Litter-Monitoring-Manual_Final.pdf

Bergmann, M., Gutow, L., Klages, M. (2015). Marine anthropogenic litter. https://doi.org/10.1007/978-3-319-16510-3.

Boucher, J., Billard, G., Simeone, E., & Sousa, J. (2020). *The marine plastic footprint*. Gland, Switzerland: IUCN. Viii + 69 pp.

BRS. (2013). Framework for the environmentally sound management of hazardous wastes and other wastes (UNEP/CHW.11/3/Add.1/Rev.1). http://www.basel.int/Implementation/CountryLedInitiative/EnvironmentallySoundManagement/ESMFramework/tabid/3616/Default.aspx

BRS. (2019a). Revised draft practical manuals on extended producer responsibility and financing systems for environmentally sound management (UNEP/CHW.14/5/Add.1). http://www.basel.int/Implementation/Plasticwaste/Guidance/tabid/8333/Default.aspx

BRS. (2019b). Revised draft guidance to assist Parties in developing efficient strategies for achieving recycling and recovery of hazardous and other wastes (UNEP/CHW.14/INF/7). http://www.basel.int/Implementation/Plasticwaste/Guidance/tabid/8333/Default.aspx

BRS. (2019c). Report on the activities of the Basel and Stockholm conventions regional centres addendum plastic and toxic additives, and the circular economy: The role of the Basel and Stockholm Conventions (UNEP/CHW.14/INF/29/Add.1, UNEP/POPS/COP .9/INF/28/Add.1). http://www.brsmeas.org/Default.aspx?tabid=7832

Bruce-Vanderpuije, P., Megson, D., Reiner, E. J., Bradley, L., Adu-Kumi, S., & Gardella Jr, J. A. (2019). The state of POPs in Ghana- A review on persistent organic pollutants: Environmental and human exposure. *Environmental Pollution, 245*, 331–342.

CBD. (2021). Preparations for the Post-2020 Biodiversity Framework. https://www.cbd.int/conferences/post2020

CIEL. (2019a). *Plastic and climate: The hidden costs of a plastic planet*. www.ciel.org/plasticandclimate

CIEL. (2019b). *Plastic and climate: The hidden costs of a plastic planet*. https://www.ciel.org/wp-content/uploads/2019b/02/Plastic-and-Health-The-Hidden-Costs-of-a-Plastic-Planet-February-2019b.pdf

City of Cape Town, Williams, S., Crous, M., & Ryneveldt, L. (2019). *Economic performance indicators for Cape Town*. http://www.capetown.gov.za/work%20and%20business/doing-business-in-the-city/business-support-and-guidance/economic-reports/Economic%20resources%20and%20publication

Climate Legal. (2020). *Policy effectiveness assessment of selected tools for addressing marine plastic pollution*. Extended producer responsibility in South Africa. Bonn, Germany: IUCN. Environmental Law Centre, 19.

DEFF. (2017). South Africa's oceans economy. https://www.gov.za/sites/default/files/gcis_document/201706/saoceaneconomya.pdf

DEFF (2020a). National Wase Management Strategy 2020a. https://www.environment.gov.za/sites/default/files/docs/2020nationalwaste_managementstrategy1.pdf

DEFF. (2020b). *Waste picker integration guideline for South Africa: Building the recycling economy and improving livelihoods through integration of the informal sector*. DEFF and DST: Pretoria.

Department of Environmental Affairs of the Republic of South Africa. (2019). Draft 2019 revised and updated national waste management strategy. *42879*, 44–45. https://www.environment.gov.za/sites/default/files/gazetted_notices/nemwa_wastestrategyrevised_g42879gon1561.pdf

Dunlop, S. W., Dunlop, B. J., & Brown M. (2020). Plastic pollution in paradise: Daily accumulation rates of marine litter on Cousine Island, Seychelles. *Marine Pollution Bulletin, 151*, 110803. https://doi.org/10.1016/j.marpolbul.2019.110803

Excell, C., Salcedo-La Viña, C., Worker, J., & Moses, E. (2018). *Legal limits on single-use plastics and microplastics: A global review of national laws and regulations*. UNEP. https://www.unep.org/resources/report/legal-limits-single-use-plastics-and-microplastics

GAIA. (2020). *Waste pickers hold skill exchange in Kenya*. https://www.no-burn.org/waste-picker-exchange-in-kenya

Geyer, R., Jambeck, J. R., & Law, K. L. (2017). Production, use, and fate of all plastics ever made. *Science Advances, 3*(7), 25–29. https://doi.org/10.1126/sciadv.1700782

Gilman, E. (2015). Status of international monitoring and management of abandoned, lost and discarded fishing gear and ghost fishing. *Marine Policy, 60*, 225–239.

Global Ghost Gear Initiative (2020). *Sofer initiative—Fishing net gains Nigeria*. https://www.ghostgear.org/projects/sofer-initiative

Goddard, H. C. (1995). The benefits and costs of alternative solid waste management policies. *Resources, Conservation and Recycling, 13*(3–4), 183–213. https://www.sciencedirect.com/science/article/abs/pii/092134499400021V

Godfrey, L. (2019). Waste plastic, the challenge facing developing countries—Ban it, change it, collect it? *Recycling, 4*. https://doi.org/10.3390/recycling4010003

Guardian. (2020). *Oil industry lobbies US to help weaken Kenya's strong stance on plastic waste*. Guardian News & Media. https://www.theguardian.com/world/2020/sep/01/kenya-plastic-oil-industry-lobbies-us

IMO. (2018). *Addressing marine plastic litter from ships—action plan adopted*. Addressing marine plastic litter from ships—action plan adopted (imo.org).

Iñiguez, M. E. & Conesa, J. A., & Fullana, A. (2017). Microplastics in spanish table salt. *Scientific Reports, 7*. https://doi.org/10.1038/s41598-017-09128-x

IUCN-EA-QUANTIS (2020a). *National Guidance for plastic pollution hotspotting and shaping action*, Country report Kenya. https://plastichotspotting.lifecycleinitiative.org/wp-content/uploads/2020a/12/kenya_final_report_2020a.pdf

IUCN-EA-QUANTIS (2020b). *National Guidance for plastic pollution hotspotting and shaping action, Country report Mozambique*. https://plastichotspotting.lifecycleinitiative.org/wp-content/uploads/2020b/12/mozambique_final_report_2020b.pdf

IUCN-EA-QUANTIS (2020c). *National Guidance for plastic pollution hotspotting and shaping action, Country report South Africa (updated)*. https://plastichotspotting.lifecycleinitiative.org/wp-content/uploads/2021/05/SouthAfrica_final_report_2020c_UPDATED.pdf

IUCN-EA-QUANTIS (2020d). *National Guidance for plastic pollution hotspotting and shaping action, Country report Tanzania*. https://plastichotspotting.lifecycleinitiative.org/wp-content/uploads/2021/05/Tanzania_final_report_2021.pdf

Jain, A., Raes, L., & Manyara, P. (2021). *Efficiency of beach clean-ups and deposit refund schemes (DRS) to avoid damages from plastic pollution on the tourism sector in Cape Town, South Africa*. IUCN. 10, Switzerland. https://www.iucn.org/sites/dev/files/content/documents/marplasticcs_economic_policy_brief_south_africa_final.pdf

Jambeck, J., Hardesty, B. D., Brooks, A. L., Friend, T., Teleki, K., Fabres, J., et al. (2018). Challenges and emerging solutions to the land-based plastic waste issue in Africa. *Marine Policy, 96*, 256–263. https://doi.org/10.1016/j.marpol.2017.10.041

Kobo, K. (2014). *Ivory Coast defiant on plastic bags ban, traders upset.* Anadolu Agency. https://www.aa.com.tr/en/life/ivory-coast-defiant-on-plastic-bags-ban-traders-upset/165144

Knox, J.H. (2020). Constructing the human right to a healthy environment. *Annual Review of Law and Social Science, 16*, 79–95. https://doi.org/10.1146/annurev-lawsocsci-031720-074856

Ladan, M. T. (2018). Achieving sustainable development goals through effective domestic laws and policies on environment and climate change. *Environmental Policy and Law; Amsterdam, 48*, 42–63. https://doi.org/10.3233/EPL-180049

Larsen, J., & Venkova, S. (2017). *The downfall of the plastic bag: A global picture.* Earth Policy Institute. https://earthpolicyinstitute.wordpress.com/page/2/

Lebreton, L., & Andrady, A. (2019). Future scenarios of global plastic waste generation and disposal. *Palgrave Communications, 5*(1), 1–11. https://www.nature.com/articles/s41599-018-0212-7

Masron, T. A., & Subramaniam, Y. (2019). Does Poverty cause environmental degradation? Evidence from developing countries. *Journal of Poverty, 23*, 44–64. https://doi.org/10.1080/10875549.2018.1500969

McIlgorm, A., Raubenheimer K., & McIlgorm, D. E. (2020). *Update of 2009 APEC report on Economic Costs of Marine litter to APEC Economies.* A report to the APEC Ocean and Fisheries Working Group by the Australian National Centre for Ocean Resources and Security (ANCORS), University of Wollongong, Australia. https://www.apec.org/docs/default-source/Publications/2020/3/Update-of-2009-APEC-Report-on-Economic-Costs-of-Marine-Debris-to-APEC-Economies/220_OFWG_Update-of-2009-APEC-Report-on-Economic-Costs-of-Marine-Debris-to-APEC-Economies.pdf

New York Times (2020). *Big oil is in trouble. Its plan.* Flood Africa with Plastic. https://www.nytimes.com/2020/08/30/climate/oil-kenya-africa-plastics-trade.html

Obi, I. (2009). African Circle invests $43m to reduce ship pollution in Lagos. https://www.vanguardngr.com/2009/08/african-circle-invests-43m-to-reduce-ship-pollution-in-lagos/

OECD. (2016). Extended producer responsibility: Updated guidance for efficient waste management. *OECD Publishing, Paris.* https://doi.org/10.1787/9789264256385-en

OECD. (2018). Improving markets for recycled plastics: Trends. *Prospects and Policy Responses.* https://doi.org/10.1787/9789264301016-en

Oelofse, S. H. H. & Godfrey, L. (2008). Towards improved waste management services by local government—A waste governance perspective. In: *Proceedings of the 2nd CSIR Biennial Conference*, 17–18 November 2008, Pretoria, South Africa. http://playpen.meraka.csir.co.za/~acdc/education/CSIR%20conference%202008/Proceedings/CPA-0002.pdf

Okonkwo, T. (2017). Maritime boundaries delimitation and dispute resolution in Africa. *Beijing Law Review, 8*, 55. https://doi.org/10.4236/blr.2017.81005

Olapido, D. Niang, M. (2021). *Why Senegalese women are protesting a ban on plastic.* Bloomber CityLab and Equality. Available at: https://www.bloomberg.com/news/articles/2021-11-29/why-women-in-senegal-are-protesting-a-ban-on-plastic

Opondo, G. (2020). *Policy effectiveness assessment of selected tools for addressing marine plastic pollution.* Extended Producer Responsibility in Kenya. Bonn, Germany: IUCN Environmental Law Centre, 18.

Ouguergouz, F. (1993). The Bamako convention on hazardous waste: A new step in the development of the African International Environmental Law. *African Yearbook of International Law Online/Annuaire Africain De Droit International Online, 1*, 195–213. https://doi.org/10.1163/221161793X00107

Parliamentary Monitoring Group. (2008). *The Africa Institute for Environmentally Sound Management of Hazardous & Other Wastes.* Environment Department briefing. Meeting report. https://pmg.org.za/committee-meeting/9165/

Parliamentary Monitoring Group. (2014). *Questions and replies: Environmental affairs.* Internal Question Paper No. 21 of 2014. https://pmg.org.za/question_reply/512/

PERSGA/UNE (2018). *Regional Action Plan for the Sustainable Management of Marine Litter in the Red Sea and Gulf of Aden.* Report Number RP.0091. PERSGA, Jeddah, Saudi Arabia. http://persga.org/public/library/16081912898553.pdf

PETCO (2019a). *Review of PETCO Activities.* https://petco.co.za/wp-content/uploads/2020/07/PETCO-2019a-Annual-Review_FINAL.pdf

PETCO (2019b). *Designing for the environment.* https://petco.co.za/wp-content/uploads/2019b/08/PETCO_Design-for-Recyclability_Guideline-Document_2019b_FINAL.pdf

PETCO (2021). *PETCO'S How-to Guide to Section 18 for Producers.* https://petco.co.za/wp-content/uploads/2021/06/Section-18_PETCOS-HOW-TO-GUIDE-TO-SECTION-18-FOR-PRODUCERS_FINAL.pdf

PETCO Kenya (2021). *Who we are.* https://www.petco.co.ke/

Pew Charitable Trusts and SYSTEMIQ. (2020). Breaking the plastic wave. https://www.systemiq.earth/breakingtheplasticwave/

Raubenheimer, K., & Urho, N. (2020). *Possible elements of a new global agreement to prevent plastic pollution.* Nordic Council of Ministers, Copenhagen. https://www.norden.org/en/publication/possible-elements-new-global-agreement-prevent-plastic-pollution

Resource Recycling. (2019). *Basel changes may have 'bigger impact' than China ban.* https://resource-recycling.com/recycling/2019/05/14/basel-changes-may-have-bigger-impact-than-china-ban/

Russo, V., Stafford, W., Nahman, A., De Lange, W., Muniyasamy, S., & Haywood, L. (2020). *Comparing grocery carrier bags in South Africa from an environmental and socio-economic perspective.* Waste Research Development and Innovation Roadmap Research Report CSIR. https://wasteroadmap.co.za/wp-content/uploads/2020/05/22-CSIR-Final-LCSA_Bags_Final-Report-vs2.pdf

Ryan, P. G. (2020). The transport and fate of marine plastics in South Africa and adjacent oceans. *South African Journal of Science, 116,* 7677. https://doi.org/10.17159/sajs.2020/7677SADC

Ryan, P. G., Dilley, B. J., Ronconi, R. A. & Connan, M. (2019). Rapid increase in Asian bottles in the South Atlantic Ocean indicates major debris inputs from ships. *Proceedings of the National Academy of Sciences USA, 116,* 20892–20897.

Ryan, P. G., Weideman, E. A., Perold, V., Hofmeyr, G. J. G., & Connan, M. L. (2021a). Message in a bottle: Assessing the sources and origins of beach litter to tackle marine pollution. *Environmental Pollution, 288,* 117729. https://doi.org/10.1016/j.envpol.2021.117729

Southern African Development Community (SADC). (2020). *Regional Indicative Strategic Development Plan (RISDP) 2020–2030.* Gaborone, Botswana. https://www.sadc.int/files/4716/1434/6113/RISDP_2020-2030_F.pdf

SADC (2021). *Waste Management.* https://www.sadc.int/themes/environment-sustainable-development/waste-management

SAICM (2021). *Launch of the Beyond 2020 process.* http://www.saicm.org/Resources/SAICMStories/LaunchoftheBeyond2020process/tabid/5530/Default.aspx

Schluep, M., Terekhova, T., Manhart, A., Muller, E., Rochat, D. & Osibanjo, O. (2012). Where are WEEE in Africa? *Electronics Goes Green 2012+, ECG 2012 - Joint International Conference and Exhibition, Proceedings* (pp. 1–6).

Shigwedha, A. (2019). *Getting to know the Abidjan Convention.* New Era Live. https://neweralive.na/posts/getting-to-know-the-abidjan-convention

Simon, N., Raubenheimer, K., Urho, N., Unger, S., Azoulay, D., Farrelly, T., et al. (2021). A binding global agreement to address the life cycle of plastics. *Science, 373,* 43–47.

Surbun, V. (2021). *Africa's combined exclusive maritime zone concept.* Africa Report 32. Institute for Security Studies. https://issafrica.org/research/africa-report/africas-combined-exclusive-maritime-zone-concept

Takunda, T. Y., & von Blottnitz, H. (2019). Accumulation and characteristics of plastic debris along five beaches in Cape Town. *Marine Pollution Bulletin, 138,* 451–457. https://doi.org/10.1016/j.marpolbul.2018.11.065

UN (2015). Transforming our world: the 2030 Agenda for Sustainable Development. https://sdgs.un.org/2030agenda.

UNEP (n.d.). About AMCEN. https://www.unep.org/regions/africa/african-ministerial-conference-environment/about-amcen

UNEP (2017a). *Combating marine plastic litter and microplastics: An assessment of the effectiveness of relevant international, regional and subregional governance strategies and approaches.* UNEP, Nairobi. http://wedocs.unep.org/bitstream/handle/20.500.11822/21854/UNEA-3%20MPL%20Assessment-Final-2017aOct05%20UNEDITED_adjusted.docx?sequence=1&isAllowed=y

UNEP (2017b). *Marine litter socio economic study.* United Nations Environment Programme, Nairobi. https://wedocs.unep.org/bitstream/handle/20.500.11822/26014/Marinelitter_socioeco_study.pdf?sequence%20%20(page%2087)%20%E2%80%93%20MARPOL%20national%20implementation%20in%20Nigeria

UNEP (2018a). *Single-use plastic: A roadmap for sustainability.* United Nations Environment Programme, Nairobi. https://wedocs.unep.org/handle/20.500.11822/25496;jsessionid=A7FC97622AFB7F0E37F72A6F00572857

UNEP (2018b). *Africa waste management outlook.* United Nations Environment Programme, Nairobi, Kenya. ISBN No: 978-92-807-3704-2.

UNEP (2018c). *How Smuggling threatens to undermine Kenya's plastic bag ban.* https://www.unep.org/news-and-stories/story/how-smuggling-threatens-undermine-kenyas-plastic-bag-ban

UNEP (2020). National guidance for plastic pollution hotspotting and shaping action. In J. Boucher, M. Zgola, et al. (Eds.), *Introduction report.* United Nations Environment Programme. Nairobi, Kenya. https://plastichotspotting.lifecycleinitiative.org/wp-content/uploads/2020/07/National-Guidance-for-Plastic-Hotspotting-and-Shaping-Action-Final-Version-2.pdf

UNEP (2021a). *Global action to protect the marine environment from land-based pollution.* https://www.unep.org/explore-topics/oceans-seas/what-we-do/addressing-land-based-pollution/global-action-protect-marine

UNEP (2021b). *Making Peace with Nature: A scientific blueprint to tackle the climate, biodiversity and pollution emergencies.* https://www.unep.org/resources/making-peace-nature

UNEP (2021c). *Abidjan Convention workshops.* Global Partnership on Marine Litter. https://www.gpmarinelitter.org/news/news/register-abidjan-convention-workshops

US EPA. (2020). *Best practices for solid waste management: A guide for decision-makers in developing countries (EPA 530-R-20–002).* https://www.epa.gov/sites/default/files/2020-10/documents/master_swmg_10-20-20_0.pdf

van Os, E., & de Kock, L. (2021). *Plastics: From recycling to (post-consumer) recyclate: Industry views on barriers and opportunities in South Africa.* WWF South Africa, Cape Town. https://www.wwf.org.za/our_research/publications/?34562/plastics-from-recycling-to-post-consumer-recyclate

Wagner, S. & Schlummer, M. (2020). Legacy additives in a circular economy of plastics: Current dilemma, policy analysis, and emerging countermeasures. *Resources, Conservation and Recycling, 158,* 104800. https://doi.org/10.1016/j.resconrec.2020.104800

Williams, A. T., & Rangel-Buitrago, N. (2019). Marine litter: Solutions for a major environmental problem. *Journal of Coastal Research, 35*(3), 648–663. Coconut Creek (Florida), ISSN 0749-0208.

Chapter 5
The Way Forward, Building Up from On-The-Ground Innovation

Thomas Maes and Fiona Preston-Whyte

Summary This chapter of the African Marine Litter Outlook summarises the previous chapters, their findings, suggestions, and identified barriers to tackling marine litter in Africa. The importance of innovative ground-up solutions tackling waste management across Africa are highlighted in this chapter. The forward approach is then outlined through recommendations. The recommendations are covered in 10 points: 9 of which focus on local sources, with a 10th outlining the global need to tackle transboundary marine plastic litter, originating from sources outside of Africa's control.

Keywords Synopsis · Way forward · Policy solutions

5.1 Introduction

The current average per capita waste production in Africa, not taking into account waste imports, is much lower than the global average (0.78 and 1.24 kg per day, respectively) (Scarlat et al., 2015; UNEP, 2018). Despite this more conservative generation of waste, Africa lacks the infrastructure and service delivery to adequately deal with its current waste production (UNEP, 2018). Across African nations, the waste management sector is underprioritised and lacks investment; the existing infrastructure is poorly maintained and is not being upgraded. Tackling marine litter from a purely waste management perspective is thus unlikely to work in Africa. Although waste management is crucial, waste prevention or waste minimisation should occur alongside. Waste minimisation involves upstream measures, therefore reducing the amount of waste and consequently its management burden. Combining waste prevention attitudes whilst improving management is likely to be a more cost-effective approach and of particular importance to Africa as increasing economic growth (3.7% continental averaged pre-COVID-19 growth) (International Monetary Fund, 2021; United Nations,

T. Maes (✉) · F. Preston-Whyte
GRID-Arendal, Teaterplassen 3, N-4836 Arendal, Norway
e-mail: thomas.maes@grida.no

© The Author(s) 2023
T. Maes and F. Preston-Whyte (eds.), *The African Marine Litter Outlook*,
https://doi.org/10.1007/978-3-031-08626-7_5

2020) and a rapidly growing population (3.5% annual growth (UNEP, 2018; Wilson et al., 2015) is resulting in an acceleration of the overall waste production and per capita waste production. The combination of increasing economic and population growth together with insufficient waste management systems means that Africa is likely to become an escalating source of marine litter, which needs an urgent and adequate response via action planning, infrastructural and financial support (Jambeck et al., 2018; UNEP, 2018). Such response should consider existing, innovative, and successful initiatives set up by the informal and formal sector, small and medium enterprises, and Non-Governmental Organisations (NGOs), which have provided successful responses in the absence of political will or financial support. These existing projects and businesses need to be upscaled in an enabling environment to serve as best practice examples whilst providing future capacity across the African continent.

Implementation and enforcement of legal and policy frameworks regarding waste management remain an issue within Africa, either through limited capacity or limited political will. Nevertheless, where legislation and infrastructure and/or enforcement is lacking, on-the-ground innovative, practical, and cost-effective solutions have been launched. They are driven by a wide range of stakeholders, including informal/formal industries, small and medium enterprises, and NGOs. The best-acknowledged examples are the support of the informal waste pickers for the formal recycling industry and the drive by the formal sector to work with governments or within communities. The innovative industries around reuse and repurpose (sometimes supplied by the informal sector) mostly operate on a local level. As the digital footprint of most informal workers, smaller enterprises, and smaller NGOs is lacking, they are often missed in a broader analysis. However, these existing solutions have been vital in reducing the burden on governments and communities and creating both micro-and macro-economically viable solutions. The sharing of existing systems, upscaling of viable solutions, and continuous support in an enabling environment–institutionally, legally, and policy-wise–are important factors to move forward in tackling marine litter in Africa. Solutions or actions within Africa should integrate an awareness component to support grassroots activities or existing work by governments, industry (both informal and formal sectors) and NGOs. Furthermore, job creation, businesses development, through the recovery of valuable recyclables, and substantial prospects for enhancing livelihoods can be supplied by the waste sector in Africa.

Actions to tackle marine litter are dependent on the litter sources. It should be noted that land-based sources should be tackled on land. For sea-based sources of marine litter, land-based actions (such as adequate and feasible port reception facilities for waste disposal) also play an important role. For several local sea-based sources of marine litter e.g., small scale fisheries, direct action is also needed on land; however, for offshore inputs (from shipping, largely external to Africa) as well as long-distance drift, especially from south-east Asia, as shown by Ryan (2020a) and Ryan et al. (2021) the source is outside the control of Africa. For these transboundary sources, actions are rather required on a global scale.

5.2   Summary of Findings, Suggestions, and Barriers Identified in Previous Chapters

The previous chapters summarised existing information. A summary of these findings is provided below:

- Although the African plastic waste footprint is relatively low in comparison with other continents, marine litter in Africa is a current and rapidly increasing problem, with important implications for the Blue Economy (freshwater and marine) and climate change mitigation (Chap. 1).
- There is a need to address the problem of marine litter with innovative measures (Chaps. 2, 3 and 4), both waste minimisation and management will be required to tackle the issue.
- Overall, there has been limited research on marine litter across the African continent. Most research has been conducted in South Africa, with few studies from other African coastal nations (Chaps. 2 and 3). Although more research is urgently required, enough is known to acknowledge the scale of the problem, to seek solutions, and implement change now.
  - Most data on marine plastic litter focuses on the distribution, characteristics, and sources. These are often limited snapshots that do not consider temporal and other variability (Chap. 2).
  - Very little research has been done to ascertain litter's biological and ecological effects in Africa (Chap. 3).
  - The social and economic impacts are particularly not well documented (Chap. 3).
- There is a misalignment between scientific reporting, solutions, and policy implementation (Chap. 3).
- The review of international and regional legal and policy frameworks shows that the obligation to prevent marine litter from land- and sea-based sources has been established. Less clear, however, is the responsibility to provide sustainable funding for such purposes whilst ensuring a safe and healthy environment and access to tenable livelihoods (Chap. 4).
- The drivers of marine litter in Africa are complex, supporting the need to incorporate a broader range of measures and stakeholders than those included in global and regional frameworks (also referred to as instruments) for the prevention of pollution, management of chemicals and waste, and the protection of species and biodiversity (Chap. 4).
- Public awareness (including environmental education and outreach), consumer behaviour, and industry engagement play a key role in preventing marine litter and must be strengthened across Africa (Chap. 4). It is noted that awareness and education alone are not enough in many cases. Stronger incentives and/or disincentives are needed to drive behaviour change.

- Progress has been made in adopting regulations to reduce problematic plastic items, particularly through adopting plastic bag bans. Enforcement of the law– i.e., legislation as well as regulatory or administrative measures–has remained a challenge for most African countries and, where plastic bag bans have been adopted, illegal trade from neighbouring countries where no ban is in place has sometimes reduced the outcomes of these measures (Chap. 4).
- The use of Extended Producer Responsibility (EPR) schemes to fund waste management infrastructure and services is extremely limited across the continent. Such schemes must consider the effect on the informal waste sector. Improving livelihoods and poverty reduction through the creation of "green jobs" should be a key driver for improving waste management across Africa (Chap. 4).
- There are several existing fora and initiatives to prevent, reduce, or combat marine litter, coordinated at regional and sub-regional levels, including but not limited to the African Union bodies and Regional Seas programmes. However, an overarching coherent and harmonised continent-wide approach is lacking. Most of the existing initiatives cover a broad spectrum of strategies and implementation plans, such as the Blue Economy, Circular Economy, and plastic pollution more broadly, which touch on, but are not primarily focussed on marine litter prevention and reduction, nor are the efforts consolidated. The few marine litter-focussed initiatives are geographically concentrated on coastal regions, neglecting the involvement of landlocked states (Chap. 4).
- Several international legal instruments—i.e., legally binding international treaties or conventions—have been implemented into national law as necessary and, in some cases have had direct effect or primacy as soon as ratified. However, there are shortcomings in their enforcement at the national level. These shortcomings are mainly due to a lack of resources and/or of capacity. International mechanisms, such as the new Global Treaty to End Plastic Pollution, are meant to offer capacity support; however, this support may be insufficient or not fulfilled until quite some time. Some instruments have been recently amended, such as the Basel Convention concerning plastic waste. Still, there has not been enough time for such amendments to be adequately implemented at the national level to properly address plastic waste issues (Chap. 4).

The previous chapters highlighted important findings and knowledge gaps about marine litter in Africa. Based on these knowledge gaps, important chapter-specific suggestions about marine litter in Africa are outlined below:

- From a mitigation perspective:
  - Comparable datasets and baselines, combined with long-term monitoring studies, are required across the continent to measure the change in the state of the environment/leakage and mitigation effectiveness (Chap. 2).

- To support this, knowledge transfer and capacity building in certain areas of expertise is required e.g., polymer identification (Chap. 2).

- From a research perspective:

  - There is a need for more field studies quantifying litter inputs across the different size ranges to facilitate more effective interventions targeting different sources (Chap. 2). As a priority, as > 99% of the mass of plastics comes from the macro scale, to develop mitigation and measuring actions, research is needed primarily at the macro scale.
  - To provide a more robust understanding of leakages and its drivers, studies should be encouraged to compare the rate of accumulation to the rate of waste generation (Chap. 2).
  - More studies on distribution and underlying mechanisms (e.g., burial, transport, and fragmentation processes), specific to African conditions, are required (Chap. 2).
  - Despite current efforts, greater effort is needed across Africa to understand the broad spectrum of waste plastic impacts, including effects on human health, environment and ecosystems, economic implications, and social factors (Chap. 3).
  - To strengthen evidence-based policy interventions, additional studies and models to better understand the drivers for abundance, distribution, pathways and sinks of plastic pollution in the environment at scale, and underlying mass balance processes are required (Chap. 2).
  - Although there appears to be a solid foundation on distribution and sources research, there needs to be a more concerted effort to synchronise work and compatibility between studies to better understand multi-national, transboundary environments. This will assist with continent-wide solutions. Although existing studies do not necessarily focus on the drivers nor impacts, these provide a foundation and a positive future trajectory for understanding the impacts of marine litter in Africa (Chap. 3) and monitoring mitigation measures (Chap. 2).
  - Coordinated research efforts will further help to standardise sampling and data collection (Chaps. 2 and 3). Research is still conducted in silos, even amongst researchers in the same field. More workshops and fora for researchers across Africa are needed.

- From a science-policy interface perspective:

  - More cross-field engagement is needed regarding planning to mitigate the effects of marine litter (e.g., between researchers, law, and policymakers) (Chap. 3).
  - There is a need for a central database nationally and regionally to use all research efforts for decision-making purposes.
  - Continent-wide data collection, monitoring, and reporting may assist in developing a continent-wide, dedicated approach to tackling marine litter (Chaps. 2, 3 and 4).

- – Evidence-based policy is vital for countries that can little afford to deal with unintended consequences of legislation. Sharing best practices in legal and policy measures that include stakeholder engagement, design, implementation, and enforcement could provide valuable insights from African and other countries. This is particularly important for African countries that struggle to raise funds for financing waste management services and infrastructure and as such could benefit from experiences in other countries where EPR schemes have been adopted (Chap. 4).

- • From a policy perspective:

  - – Strengthening the social outcomes of policies to improve the living conditions of those most impacted by the accumulation of waste in the environment and those who work in hazardous conditions, amongst others, can provide co-benefits for society and the environment whilst working towards achieving several SDGs (Chap. 4).
  - – Consolidation of existing initiatives, action plans, and resources are needed (Chap. 4).
  - – The advantages of the inclusion of marine litter interventions in the Blue Economy and Circular Economy strategies provide a focus on socio-economic development and sustainable livelihoods, whilst more can be done to elevate the need for marine litter/pollution interventions in sub-regional economic development community strategies and action plans (Chap. 4).
  - – The region needs to endorse the new global plastic treaty aimed at eliminating all discharges of plastic into the marine environment. It is expected to present a legally binding instrument, which would reflect diverse alternatives to address the full lifecycle of plastics, the design of reusable and recyclable products and materials, and the need for enhanced international collaboration to facilitate access to technology, capacity building and scientific and technical cooperation. However, such an agreement would require behaviour change across virtually the entire population. To be effective, there is a need to strengthen implementation of related action, including through adequate and sustainable financial support, transfer of technology, and capacity building (Chap. 4).
  - – Governments and businesses across the value chain will need to shift away from single-use plastics, as well as to mobilise private finance and remove barriers to investments in research and in a new circular economy.
  - – The Basel Convention, Stockholm Conventions, and other relevant regional and international instruments can play an important role, such as in sharing information, building capacity, etc. (Chap. 4).

The previous chapters of this report identified several important barriers to dealing with marine litter in Africa:

- • There is a lack of sustainable funding mechanisms for mitigation (Chap. 4).

- There is a lack of sustainable funding for implementation and enforcement (Chap. 4).
- Although African countries are signatories to international agreements, Africa has no platform to centralise strategies/protocols that are supposed to be implemented. Coordination and centralisation should reduce unnecessary replication of work and so reduce required funding.
- Research and policy decision funding is centred on localised proposals which are sometimes not aligned to national and international needs.
- Providing the evidence to inform law and policy design requires data of long-term temporal and wide geographic scale.
- The linkage of marine litter interventions and related socio-economic benefits should be highlighted in socio-economic fora and platforms such as the sub-regional economic communities (Chap. 4). A combination of lack of awareness on the numerous opportunities presented by international instruments, mechanisms, and initiatives, including their recent changes and amendments, and a general lack of prioritisation and/or resources to act (Chap. 4).

It is noted, both in this body of work and previous, that enough knowledge exists on marine litter, both globally and in Africa, to act now and drive mitigation measures. As such, resources need to go into reduction and prevention of leakage, with scientific data measuring the effectiveness of such measures.

5.3 Discussion of Report Findings

Data availability in Africa is generally poor, and the continent's contribution to the overall global scientific knowledge base was estimated at 2.8% in 2020 (Diop & Asongu, 2021). Taking into account country wealth and comparing percentages of GDP invested in research and development in 2018, African countries are not highly ranked, with Egypt appearing highest on the list—number 38 at 0.72% of GDP, compared to the global average of 1.17% of GDP invested in research and development (The Global Economy, 2021). Nevertheless, with effective use of resources and targeted studies in relation to marine litter in Africa, as outlined in Chapters 2 and 3, a substantial foundation of key knowledge on marine litter has been developed. Importantly, enough is known *"about the impacts on marine systems to justify implementing policies to reduce the leakage of waste plastic into the environment, certainly enough to start implementing mitigation measures now"* (Ryan et al., 2020a, 2020b). Methods and best practices will need to be aligned and finetuned towards the needs of the African continent. Dedicated research and monitoring will be needed to promote the development of sustainable, affordable, innovative, and cost-efficient approaches, to show effectiveness and success to funders and to fulfil policy requirements.

There is general inadequacy of waste management and infrastructure across the African continent—depending on the country, this is linked to an absence of supporting legislation or, more often, a lack of effective implementation and/or enforcement of legislation (UNEP, 2018; see Chap. 4). Implementation and enforcement of efficient waste management across Africa is exacerbated by: lack and misapplication of sustainable funding, geographical and transport challenges, educational gaps, the social stigma of working with waste, and historical disadvantages at some national and community levels, all of which delay the development of the necessary technological infrastructure to keep pace with the increasing amounts of persistent wastes such as plastics. The lack of prioritisation of funds to waste management on both national and municipal levels supports the need for waste prevention measures (reduction, circular economy, and material lifestyle approaches) and has driven existing prevention policies (primarily focusing on reduction). Despite, or maybe because of the challenges faced, Africa has been at the forefront of some innovative policy approaches. For example, South Africa was the first country in the world to introduce a plastic bag tax in 2003. It should also be acknowledged that where enforcement exists, Africa has some of the harshest law and policy enforcement measures related to plastic pollution, with Rwanda issuing up to six months jail sentences for those smuggling plastic bags in non-compliance of the country's ban (Behuria, 2021). Rwanda and Sierra Leone have also implemented requirements to communities to clean their environments regularly. Both campaigns have been shown to have early successes and are believed to be more effective in creating long-term positive behaviour change than legal punitive measures (Dessouky et al., 2016; Wilson, 1996).

Waste also holds a value, thereby creating an opportunity for job creation, illustrated by a thriving informal sector driving collection and recycling across Africa. Approached correctly, with an enabling institutional, legal, and policy environment supported by sound and credible scientific assessment and data, Africa has the potential to achieve progress, building on the existing informal networks to create a unique and innovative waste management system. There is a need for EPRs to help incentivise this process as only materials with value are collected. To prevent leakage, value (through mechanisms such as EPRs) needs to be built into all plastic items. Following the waste hierarchy, prioritising reduction strategies coupled with a circular economy approach is key to reducing the amount of waste generated and therefore needs to be appropriately managed. For the waste hierarchy to be followed successfully strong behaviour change is required, which can only occur with successful communication strategies. Concurrently, service delivery needs to increase dramatically; this can be supported by recognising the value of waste, both in mandatory and employment terms. On a policy level, a circular economy needs to be approached regionally within Africa, ensuring circularity through the necessary reduction strategies through the redesign, phasing out or elimination of unsustainable products and materials, introduction of reuse models, and regional recycling hubs. Cohesive harmonised legal and policy frameworks are needed concerning inter and intra transboundary movement of waste into and within Africa, with adequate external border laws and regulations as well as sufficient monitoring to ensure that Africa does not become the dumping ground for waste masquerading as second-hand goods from high-income countries–as is happening with e-waste (Amoyaw-Osei et al., 2011; Grant & Oteng-Ababio, 2016; GRID-Arendal, 2020; Odeyingbo et al., 2017; Maes & Preston-Whyte, 2022).

The environmental impact of mismanagement of waste in Africa adds additional stress to an environment already pressured by climate change, agriculture, urbanisation, overfishing, and invasive species. Localised waste's social and economic impact close to urban areas adds an additional strain of reduced mental and physical well-being and economic costs. Africa's Blue Economy (freshwater and marine) holds vast untapped potential for economic growth (see Chap. 1). Whilst the principles of the Blue Economy promote sustainable development and livelihoods, protecting freshwater and marine environments is also a priority principle of this approach and should be strengthened across Africa (AU-IBAR, 2019). Climate change, biodiversity loss, and other pressures such as land use already create pressure on aquatic systems. Underpinning the Blue Economy with a green approach, thus creating a Blue-Green economy, could be a sustainable outcome for Africa's increasing population.

Marine litter is often seen as an issue for countries with coastlines. However, rivers are a conduit for marine litter (Chap. 2). Given that many African rivers are transboundary (Fig. 2.1, Chap. 2), landlocked countries have a role in tackling marine litter. But more than that, regional support and action are necessary for tackling waste production and mismanagement in Africa. The porosity of borders and plans to further open internal African borders to encourage economic growth (Gordon, 2021) means that waste needs to be tackled on a regional level to prevent further transboundary issues. Furthermore, the oceans contain a value for all nations across Africa, not only coastal countries. They act as a climate regulator, a food and nutrient source, and the regional importance of the Blue Economies (both freshwater and marine) is recognised (see Chap. 1). Marine litter is an indicator of the leakage of waste into the environment. Tackling both land-based and sea-based sources of marine litter reduces litter in upstream environmental compartments such as freshwater systems, and the terrestrial environment whilst also protecting the oceans, directly impacting environmental and human health and well-being and livelihoods. Thus, marine litter in Africa is an issue for both coastal and landlocked countries. Each country is best positioned to understand its own national conditions, including its stakeholder activities, related to addressing plastic pollution.

It is acknowledged that limited foreign investments for waste management in Africa are available. Still, these are often only at a national level, and such investment (especially long-term investment) rarely reaches the industries or NGOs working on-the-ground or the institutions responsible for local services. The reasons for this are complex but can include donors' stipulations and restrictions, which are created through a high-income country understanding of the approach to waste, rather than adjusting to specific and tailored needs in different parts of Africa. Additionally, foreign investments may only cover infrastructure investment but not sustainable finance for maintenance, operational costs, nor capacity building (EPA, 2020). The new Global Plastic Treaty acknowledges the requirement for a financial mechanism to provide for the functioning of the agreement, including the possibility of a committed joint fund. Some legal obligations arising out of a new international legally binding instrument will require capacity building and technical and financial support in order to be effectively applied by developing countries and countries with economies in transition.

Marine litter can be tackled by strengthening political will. Political will is strengthened by raising awareness of the issues caused by mismanaged waste to society and the ecosystems they rely on. Such awareness needs to be raised in a variety of fora and platforms, including economic development communities. Educating law and policymakers and enforcement authorities on possible solutions and on development of financial roadmaps can help promote regulatory frameworks that incentivise private sector investment and lower their risk. Public acceptance and engagement in prevention, reduction, reusing, and recycling strategies can be better achieved through raising public awareness of the benefits of reducing waste in the environment. However, awareness is fruitless without alternative and viable behaviour choices. Investigating and implementing economic incentives and sustainable financing strategies appropriate to Africa should be made a priority.

It is important for African countries (through regional mechanisms) to stand together to prevent the entry of plastics that are not easy to recycle or repurpose after their initial use and to redesign, phase out, ban, and minimise the entry of hazardous plastics and those with toxic additives. There is a need to strengthen custom standards and procedures to minimise rampant Harmonised System (HS) miscoding of plastic products across all African countries. HS coding refers to the internationally standardised system of names and numbers to classify traded products as set out by the HS Convention (1988) developed by the World Customs Organization. HS miscoding has been highlighted as an issue in Africa regarding waste imports. In addition, there is a need to not over-emphasise conventional recycling as a viable solution for Africa but instead promote reduction and substitution of more sustainable products through appropriate legislation and behaviour. As legislation requires adequate enforcement, additional approaches using sufficient incentives should be used to affect behaviour change. Incentives such as EPR schemes add value to waste and drive behaviour change without costly enforcement of punitive measures.

There is a need for policies to address mismanaged plastic waste, uncollected waste, street littering, the thousands of illegal and unregulated dumpsites (through their recognition and formalisation), as well as stopping practices such as open burning. The success of mitigation measures, and awareness should be monitored and underpinned with scientific assessments. Cross-field and cross-border scientific collaboration are needed to support such change, with co-ordinated and comparable research methods. The African Marine Waste Network is one such network working towards this. The African Marine Litter Monitoring Manual (Barnardo and Ribbink, 2020) provides practical guidance on monitoring different environmental compartments and size fractions. However, more coordination is needed on a regional level regarding implementing of adopted actions and measuring the success of such actions through scientific research. Additionally, successful piloted actions should be scaled up using existing and new resources, ensuring that proven sustainably financed enterprises reduce and ultimately prevent marine litter.

5.3.1 A Note on the COVID-19 Pandemic and Marine Litter in Africa

Globally, the COVID-19 pandemic has led to an increase in production and consumption of single-use plastics especially personal protective equipment (Prata et al., 2020), as well as plastic packaging for food and plastic bags (Filho et al., 2021). Benson et al. (2021) estimated that over 12 billion medical and fabric face masks are discarded monthly during the pandemic in Africa, equating to 105,000 tonnes of face masks per month, which without proper management might be disposed of into the environment. COVID-19 has led to an observed increase in marine litter in Africa through higher consumption levels of COVID-19 related products (Okuku et al., 2021)—explicitly referring to the following COVID-19 related products: masks, gloves, sanitiser containers, soap wrappers, wet wipes, and liquid hand wash bottles. Research on the change of behaviour patterns during lockdowns has highlighted the importance of foot traffic to the levels of both beach (Okuku et al., 2021) and street litter (Ryan et al., 2020a).

The tough lockdowns, seen at the beginning of the pandemic in countries such as South Africa, Uganda, and Sierra Leone, and the corresponding enforced restriction on movement, were seen to have a devastating impact on waste pickers and thus negatively affected the recycling industry. For example, in Sierra Leone, the general cleaning exercises conducted between 5.00 am and 12.00 noon on the last Saturday of every month were discontinued during this COVID lockdown. However, as lockdowns were implemented with great variety across Africa (Haider et al., 2020), the impact of such measures across the continent is largely unquantified. The pandemic has negatively impacted economic growth on the continent (African Development Bank Group, 2021; Inegbedion, 2021). The COVID-19 pandemic has caused the worst economic recession in Africa in half a century (-2.1% real GDP in 2020). Though initial analysis by the African Development Bank Group (2021) expects rapid recovery in the years to follow, the United Nations (2022) predictions are more cautious showing slow recovery for Africa, below pre-pandemic predictions. Recovery is expected to be driven through the "resumption of tourism, a rebound in commodity prices, and the rollback of pandemic-induced restrictions". The outlook is, however, subject to great uncertainty from both external and domestic risks. Given this projection, though acknowledging the uncertainty of the COVID-19 pandemic in forecasts, the African Marine Litter Outlook considers planning for waste management based on future projections an absolute necessity.

Data on the short-term and long-term impacts of the COVID-19 pandemic on waste management and marine litter is currently difficult to quantify, especially in Africa. The pandemic has, however, caused a short-term shift in priorities to focus on COVID-19 relief and corresponding redirection of funds and efforts. Long-term effects on waste production through population and as a result economic growth impacts, as well as waste infrastructure investment and policy impacts are difficult to quantify.

5.4   Overall Recommendations

The Africa Waste Management Outlook (UNEP, 2018) provides in-depth recommendations for improving waste management across Africa. The authors of this African Marine Litter Outlook recognise the broad coverage of the Africa Waste Management Outlook and have focused the recommendations within the African Marine Litter Outlook on measures related specifically to marine litter, from local sources, with a touch on national and international needs. Several African groupings exist, but hardly deal with marine issues, they rather tackle terrestrial and freshwater issues e.g., The African Ministers' Council on Water (AMCOW). There are also multiple regional economic communities e.g., The Economic Community of West African States (ECOWAS), The United Nations Economic Commission for Africa (ECA), the Southern African Development Community (SADC), The Intergovernmental Authority on Development (IGAD), however most of these communities lack ocean mandates which would be relevant to tackle marine litter.

The recommendations put forth in this outlook should not be seen as static but rather should be routinely updated with an evidence-based approach, as shown in Fig. 5.1. Mitigation measures, approaches or actions should be monitored. Their effectiveness measured, and their implementation revised every few years to ensure a cohesive, efficient approach. This will ensure that ineffective measures do not continue, and new actions can be brought in as needed.

For every action, implementation should be approached to ensure that the action becomes autonomous and self-funded in the long term. The cost of funding should be borne by the polluters through schemes that add value to the waste—which in turn creates a continuous and sustainable funding mechanism whilst creating long-term behaviour change on individual and industry levels. EPR's and polluter pay's principles are two examples out of many such mechanisms. Positive incentives that drive behaviour change are more likely to drive change in an environment like Africa, where enforcement and funding are often an issue.

The recommendations outlined below are not listed in order of priority.

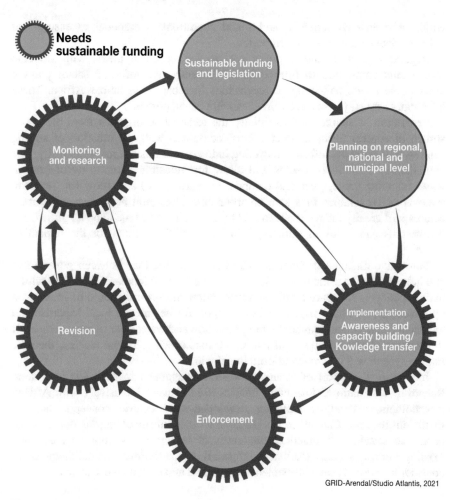

Fig. 5.1 Circular evidence-based approach

5.4.1 *Prioritise and Finance Innovative Waste Management in Africa*

The current "traditional" waste management systems employed globally (household waste collected by a local authority and recycled, incinerated, or landfilled) have shown to be mostly ineffective in Africa. However, at the grassroots level, informal systems, industry, and NGOs tend to fill some of the gaps, creating innovative waste management systems across Africa. Innovation in waste management in Africa needs to be prioritised and financially supported. There is a need to develop African-centric solutions through an enabling institutional, legal, and policy environment. Developing and sharing best practices within African environmental and social conditions is essential to ensure that inappropriate solutions are not imported.

Regular collection and cleaning are needed in urban areas, with strategic investments committed to litter control. Currently, the informal sector plays a crucial role in the collecting of recyclables in urban areas across Africa. These informally set up systems and workers should be supported.

Separation at source, and especially the removal of organics from the waste stream at source, needs to occur. This, combined with the creation of sorting centres, will assist the informal recycling industry and waste recovery industry. The combination of these two steps will allow for industrial composting of organic waste (creating value, jobs, and compost for agriculture) and allow for the clean removal of recyclables in a safe environment. Separation at source and sorting centres will greatly reduce the amount of waste that then needs to be landfilled. So, investments can occur in sanitary, well-run landfills systems for the remaining waste.

Innovative financial mechanisms can help share the load between government and industry. At the same time, a dedicated (ring-fenced) increase in funding within national budgets for marine pollution prevention and control should be introduced by governments. These should include support for heightened local expertise and technical capacity building concerning pollution and water quality management. If the sources of marine litter, and marine plastics specifically, are tackled, then the marine system will be protected from this threat.

International support of financing waste management in Africa is important. Regarding international financing, funders need to work inclusively with African stakeholders, instead of dictating developed-word centric concepts that are inefficient (or less efficient) in an African environment (including institutional and economic settings). A practical awareness of legacy issues, running costs, and existing working systems (formal or informal) is vital for international financiers to consider in order to make investments in efficient and sustainable systems.

5.4.2   Create an Enabling National Environment Through the Adoption of Adequate National Institutional, Legal and Policy Frameworks

There is a need for adequate national legislation or regulation or other legal and policy measures to enable institutions properly and to support innovation in the circular economy and waste management and allow for the development of regional support.

Multi-sectoral institutional and other mechanisms need to be strengthened and established; partnerships between relevant stakeholders need to address waste management. The broader context of national legal and policy and planning frameworks requires to integrate terrestrial and marine pollution prevention and control measures and policies.

5.4.3 Strengthen and Harmonise Existing Regional Governance to Support Cohesive Homogenised National Institutional Structures, Policies, as Well as Legislative and Regulatory Measures Aligned with International Mandates and Commitments

A cohesive, harmonised regional approach is needed concerning the transboundary movement of waste, both with regards to inter and intra movement in Africa. This needs to cover imported waste, second-hand products, and charitable donations. Harmonised, strong commitments to reduce and/or eliminate where possible the production and consumption of common and persistent litter items are needed across Africa.

Sharing knowledge and resources will save time and resources regionally. It is acknowledged that support from existing regional and international instruments (such as the Abidjan Convention, Basel Convention, Stockholm Convention, etc.) can, and should, be fully harnessed. Using existing frameworks saves resources, however, little will be achieved without focused aims, development and binding commitments focused on reducing waste formation, improving waste management, and preventing marine litter. Work within existing frameworks needs to focus specifically on preventing marine litter. The current development of a regional legal framework against plastic pollution and national marine litter action plans through the Abidjan Convention is a good example of knowledge and resource sharing.

5.4.4 Investment in Implementation and Enforcement of National, Regional, and International Legal and Policy Frameworks

The duty to prevent marine litter has been clearly established in international and regional frameworks, however, there is a lack of clarity on several key aspects in local and national implementation of regional and international commitments as subscribed under legally binding international legal instruments. Better cooperation, coordination, and collaboration of interventions between relevant stakeholders in the circular economy and the Blue Economy is also necessary. Additionally, the responsibility of obtaining sources of sustainable funding is not clear. Clarity at a local and national level is needed and investment in implementation and enforcement. Systems such as EPRs for national level law and policy can ensure accountability and financing. Other market-based instruments such as container deposit schemes may allow for the development of sustainable financing. Capacity building through training and technical assistance offered by instruments like the Basel Convention should be fully utilised.

In the best practices for developing countries, EPA (2020) outlines that "Prioritising solid waste management, researching cost-cutting strategies, incorporating pay-as-you-throw programs or taxes, and partnering with international investment organisations are all options for funding viable solid waste programs. Although some programs, taxes, or fees will face resistance when introduced, finding a sustained source of funding for solid waste management is an integral part of a successful program". Public awareness and communication are essential regarding both systems, whether requiring behaviour change or an increased cost born by citizens to pay.

5.4.5 Raise Public Awareness About the Importance of Waste Management, Water Quality, and Marine Ecosystems to Induce Behavioural Change

There is limited and ongoing need to increase awareness of the relationship between development and environmental protection. Similarly, there is limited awareness between ecosystem health and the production of ecosystem services and the Blue Economy. In addition to regular waste collection, changes in perception about the value of waste, waste mismanagement, and the environment are needed to induce positive behavioural changes regarding reduction, improper solid waste disposal, littering, separation at source and recycling. Behavioural changes regarding upstream interventions are needed to reduce plastic production, reduce waste, and support reuse (through product take-back schemes) and circularity, thus reducing waste overall. Education and awareness strengthen implementation and support existing initiatives.

Public education (including through formal education systems), awareness campaigns, and targeting specific user groups (e.g., fishers) all play an important role in minimising the impact of marine pollution. Public education plays an important role in creating support for any behavioural change needed to support policy (such as separation at source). Globally, few studies have assessed the effectiveness of education campaigns on long-term behavioural change regarding marine litter. To establish their effectiveness, education campaigns should be accompanied by studies including integrative actions and respective long-term methodological triangulation evaluations (Bettencourt et al., 2021).

5.4.6 Improve the Analytics and Knowledge Base on Marine Pollution and Water Quality Throughout the Region Using Common Monitoring Approaches and Guidelines

Africa has a scarcity of comparable quality-assured environmental data. Academics and NGOs are working to ensure the application of methods for comparable data sets on macro and micro marine litter across Africa (Barnardo and Ribbink, 2020; CEFAS, 2020). Comparable data sets, resulting from the same or equivalent methods, allow for regional and global comparisons. Such data sets are currently focused on developing baseline assessments (where lacking) and quantifying sources and hotspots. This will ensure that legal and political decisions can be based on scientific information. And can measure how efficient mitigation measures are effective at both a local and regional level. Regarding marine litter, additional comparable data sets also need to be built upstream of the environmental observations, such as household waste audits, port reception facility audits, transboundary datasets of waste or second-hand goods movement, and social economic and perception studies to understand behaviour change over time.

Monitoring efforts should be integrated into relevant regional assessments and reporting efforts, particularly the Abidjan, Nairobi, and Barcelona Conventions. For this reason, the Abidjan Convention Secretariat, in partnership with GRID-Arendal, has been working in three pilot countries (Sierra Leone, Benin, Côte d'Ivoire, and Ghana) to build capacity to develop a state of the marine environment report. Such programmes should be extended to other countries in the Abidjan Convention area. Marine litter, and corresponding data, should be integrated into SDG matrices. This will encourage cross-sector collaboration in mitigation measures. The Abidjan Convention is currently developing National Marine Litter Action Plans, of which monitoring is a part. Monitoring marine litter through earth observation is a developing field (Biermann et al., 2020) which, given its ability to track litter over large geographical ranges, Africa should consider it.

5.4.7 Measure the Economic Impacts of Marine Pollution, and Quantify the Costs Associated with Pollution Prevention and Management, as Well as the Costs Associated with Doing Nothing

The economic impacts of waste mismanagement, and the resulting pollution need to be better understood, especially in context of the Blue Economy and sustainable development. This should include clean-up costs (regular clean-up and disaster clean-ups) of streets, beaches, and ports and any economic losses in industries such as tourism and fisheries (cost of lost or abandoned gear). The social and health impacts should also be assessed to inform law, policymakers, and the public.

Industrial analysis to support regional solutions is also needed, as well as analytics on incentives, disincentives, and standards.

5.4.8 Implement Integrated, High-Priority Interventions to Reduce the Discharge of Untreated Sewage and Nutrients and Promote Wastewater Resource Recovery

Proper wastewater management is key to ensuring human and ecosystem health, economic and environmental benefits. Proper sanitation and wastewater treatment can tackle marine litter (through the removal of both macro and microplastics) and eutrophication and human health issues. The occurrence of microplastics in sludge or biosolids used in agriculture is an emerging field of research, especially in Africa (Okoffo et al., 2021). Africa's continued population and economic growth is placing pressure on the existing wastewater and stormwater drain networks—specifically in densely populated urban settlements (African Development Bank et al., 2020). From a marine litter perspective, wastewater management removes both macro litter and between 88–94% of microplastics, depending on the level of treatment (Lyare et al., 2020). Whilst significant efforts have been made across Africa to ensure better sanitation, many places still have inadequate sanitation and wastewater management (African Development Bank et al., 2020).

Nutrient enrichment of coastal and marine waters is the primary cause of eutrophication that leads to the formation of algal blooms. Eutrophication leads to hypoxic and anoxic conditions in water, extreme turbidity, and threat to marine life (Malone & Newton, 2020). Nutrient input to the marine environment can be derived from the discharge of untreated sewage and industrial/domestic wastewater into river courses. In Africa, due to the poor state of water and sanitation facilities (Yasin et al., 2010), a significant proportion of the nutrient input originates from sewage disposal. Nutrient enrichment of coastal and marine waters is the primary cause of eutrophication that leads to the formation of algal blooms. As such, eutrophication is probably a good proxy for microplastic presence and distribution in Africa, hotspots might be more readily identified by using available water quality information. Several eutrophic coastal areas now affect countries around the African continent, namely Côte d'Ivoire, Egypt, Ghana, Kenya, Mauritius, Morocco, Nigeria, Tanzania, Tunisia, Senegal, and South Africa (Diaz et al., 2011).

In Africa, viable wastewater-based resource recovery initiatives are emerging with public–private partnerships (African Development Bank et al., 2020), which follow a circular economy approach. Implementing integrated wastewater treatment improves sanitation (and associated benefits), protects freshwater resources, contributes to agriculture and energy needs, and tackles an important source of marine litter.

5.4.9 Improve Chemical and Industrial Pollution Control Through Targeted and Cost-Effective Measures in Priority Issues

Industry generates a substantial amount of wastewater. Although significant industrial hubs are limited to a few countries in Africa such as South Africa, Egypt, Morocco, and Tunisia. Mining, paper mills, tanneries, textiles, food, and beverage production, sugar refineries, oil production, and pharmaceutical production have been flagged as major contributors to the discharge of toxic wastewater (African Development Bank et al., 2020). Wastewater reuse and treatment for industry have several benefits, including pollution (including microplastics) reduction.

Eutrophication, as discussed under point 8 is driven by nutrient input into the marine environment. It is primarily derived from land-based sources, mainly through stormwater runoffs over agricultural land where nitrogen, phosphorus, and potassium (N–P–K)-based fertilisers are applied.

Chemical and industrial pollution control go further than wastewater. With adequate pollution control, marine litter, and chemical pollutants (that can be sorbed and transport by plastics), can be greatly reduced.

5.4.10 International Responses are Needed to Deal with Transboundary Waste

Within Africa, there is a strong signal that litter found close to urban centres originates from local sources (Ryan et al., 2018; Weideman et al., 2020). However, on the eastern boundary specifically, transboundary plastic litter from south-east Asia and ship sourced waste has been identified as a further source (Duhec et al., 2015; Ryan et al., 2021; Ryan, 2020b; van der Mheen et al., 2020) (see Chap. 2 for further details). Though Africa can tackle its local sources, the long-distance drift of marine litter through currents is beyond African intervention. Global solutions, such as the Global Treaty to End Plastic Pollution, are needed to support Africa in tackling international sources of marine litter.

Global solutions for reducing plastic marine litter are welcomed. These include reduction and circularity of design in all products. But more than that, a coordinated global response is necessary through this global agreement that practically reduces waste formation and stops leakage into the environment.

5.5 Steps to Consider for Local Sources

Stakeholder engagement is central to preventing marine litter through reduction, waste minimisation, reuse, recycling, and waste management. In Africa, with many stakeholders successfully operating in the reuse and recycling space specifically, stakeholder engagement becomes pivotal when planning on, and implementing changes or new concepts. Figure 5.2 indicates the stakeholders generally found in the waste management space.

First and foremost, waste minimisation through reduction, reuse and recycling should be considered. This is particularly pertinent in Africa, where funding is an issue.

Considering plastics specifically, the plastic flows and leakage assessments carried out by IUCN/UNEP in Kenya (IUCN et al., 2020a), Tanzania (IUCN et al., 2021a), Mozambique (IUCN et al., 2020b), and South Africa (IUCN et al., 2021b)

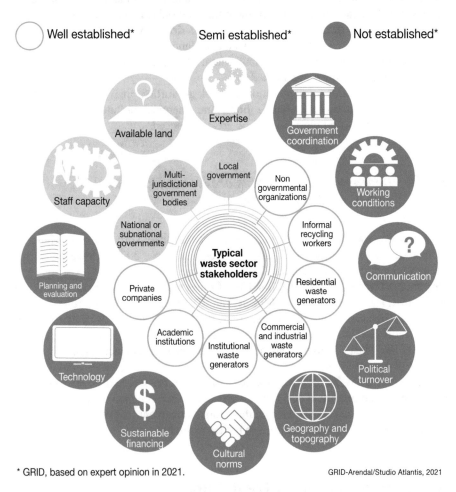

* GRID, based on expert opinion in 2021.   GRID-Arendal/Studio Atlantis, 2021

Fig. 5.2 Considerations in tackling marine litter through waste management—African stakeholder engagement, on average. Adapted from EPA (2020)

outline priority interventions across all lifecycle stages of plastics to minimise leakage potential.

Once waste is reduced, waste management planning should consider the following factors (see Fig. 5.2) (EPA, 2020):

- Social-economic factors:
  - Costs of not taking action on e.g., tourism, human health
  - Operational costs
  - Sustainable financing of implementation, enforcement, and monitoring

- Technical and staff capability:
  - Equipment and solutions that suited to environmental and social conditions
  - Technical capacity required for equipment and solutions
  - Staff capacity and expertise (or sustainable finance needed for training)

- Political changes:
  - ensuring systems and initiatives can survive administrative changes
  - establish long-term, sustainable systems that continue across administrations
  - work towards long-term staff and industry commitments

- Planning and evaluation on regional, national, and municipal levels
- Coordination between stakeholders, frameworks, and government departments
- Improving working conditions for skills retention (in the informal and formal industry)
- Stakeholder engagement (as highlighted above)—especially with existing working informal and formal sector and NGO projects
- Availability of space—including adequate space for the informal and formal private sector and government sector to work safely with economically viable distances
- Climatologic, geographic, and topographic conditions influence the availability and cost of equipment, the feasibility of technologies, and operating costs.
- Cultural norms:
  - Changing consumption and waste disposal patterns
  - Projections on waste production linked to population and economic growth

- Behavioural aspects of individual people, and the reasons behind them

5.6   Concluding Remarks

Africa has a predominantly young population. A young population is indicative of a dynamic, innovative population with huge potential to implement change. Despite limited resources, Africa has shown innovative solutions to waste management, resource recovery, and new solutions in tackling marine litter. These solutions are predominantly driven by the industry (informal and formal sectors) and NGOs. These existing solutions should be supported, and where financially sustainable, scaled up to cover new and more extensive areas (where economically viable through the economics of scale) or integrated into the implementation of adequate institutional, legal, and policy frameworks.

Given the diversity of the African continent, there is a need to develop a decision framework for local, national, and regional actions to feed into global commitments. This will assist African nations to implement the best measures for their unique social and economic situations. Each country is best placed to appreciate its own national solutions and limitations. This includes stakeholder involvement, financial and technical capacity needs related to addressing plastic pollution issues.

Mitigation actions need to target different sources of marine litter. The actions required to target local sources occur on land before the litter enters the African waters. Tackling sea-based sources, where the ships stop (or are based) at African ports, requires both sea-based (behaviour) and land-based (port reception facilities) approaches. However, to tackle offshore inputs of marine litter into Africa, both sea-based sources and transboundary sources e.g., originating from south-east Asia, Africa needs the international communities to support in implementing actions across boundaries and regions.

Regarding local sources, tackling marine litter and waste mismanagement in Africa has the potential to both create new jobs and protect existing jobs, in particular those related to the Blue Economy. By successfully dealing with this issue, Africa can contribute to better pollution control, mitigate climate change protect biodiversity, and achieve other SDGs.

Additionally, nature-positive solutions are still extensively utilised across Africa and have the potential to grow and to enhance partnerships between industry and government. Nature-positive solutions have the potential to grow and enhance the sustainable growth of the Blue-Green and circular economy, tackle climate change, mitigate climate change, improve sanitation and wastewater management, and reduce marine litter.

With limited resources, African researchers have developed substantial research, focusing on characterisations, amounts, and distributions. Even with the existing knowledge gaps, the scale of the current problem is clear. The status and future projections mean that its paramount to implement mitigation now, without waiting on further research on the scale of marine litter or impacts.

Monitoring mitigation effectiveness is needed to support the science-policy interface. The development of the science-policy interface in Africa can bring about rapid, sustainable change regarding the circular economy, waste management, and marine litter. This can occur provided a cohesive and homogenised enabling institutional, legal, and policy environment is created to support innovation and public–private sector partnerships at national and regional levels. Existing frameworks and networks within Africa can support such change, helping to implement reduction, resource recovery, and service delivery, monitored with comparable techniques to measure long-term mitigation effectiveness.

Acknowledgements We would like to acknowledge the valuable insights proved by Clever Mafuta, Morten Sorensen, Peter Ryan, Anham Salyani, Salieu Sankoh, Tony Ribbink, Abdoulaye Diagana, Alison Amoussou, as well as Jost Dittkirst and the Plastic Task Force of the Secretariat of the Basel, Rotterdam and Stockholm Conventions in peer-reviewing this chapter. We would also like to acknowledge Nieves López and Federico Labanti (Studio Atlantis) for creating the illustrations.

References

AfDB, UNEP, GRID-Arendal. (2020). *Sanitation and wastewater Atlas of Africa*. Abidjan, Nairobi and Arendal. https://www.grida.no/publications/471

African Development Bank Group. (2021). *African economic outlook 2021: From debt resolution to growth: The road ahead for Africa*.

Amoyaw-Osei, Y., Agyekum, O. O., Pwamang, J. A., Mueller, E., Fasko, R., Schluep, M. (2011). *Ghana e-waste country assessment*. SBC E-Waste Africa Project.

AU-IBAR. (2019). *Africa blue economy strategy*. https://www.au-ibar.org/sites/default/files/2020-10/sd_20200313_africa_blue_economy_strategy_en.pdf

Barnardo, T., & Ribbink, A. J. (2020). *African marine litter monitoring manual*. African Marine Waste Network, Sustainable Seas Trust. https://repository.oceanbestpractices.org/handle/11329/1420

Behuria, P. (2021). Ban the (plastic) bag? Explaining variation in the implementation of plastic bag bans in Rwanda, Kenya and Uganda. *Environment and Planning C: Politics and Space*. https://doi.org/10.1177/2399654421994836

Benson, N. U., Fred-Ahmadu, O. H., Bassey, D. E., & Atayero, A. A. (2021). COVID-19 pandemic and emerging plastic-based personal protective equipment waste pollution and management in Africa. *Journal of Environmental Chemical Engineering, 9*, 105222. https://doi.org/10.1016/j.jece.2021.105222

Bettencourt, S., Costa, S., Caeiro, S. (2021). Marine litter: A review of educative interventions. *Marine Pollution Bulletin, 168*. https://doi.org/10.1016/j.marpolbul.2021.112446

Biermann, L., Clewley, D., Martinez-Vicente, V., Topouzelis, K. (2020). Finding plastic patches in coastal waters using optical satellite data. *Scientific Reports, 10*. https://doi.org/10.1038/s41598-020-62298-z

Cefas (Centre for Environment, F.& A.S., 2020. *CSIR—DSI—Cefas Marine plastic litter workshop Day 2—Report*.

Dessouky, N., Moustafa, Y., Tutwiler, R., Estevez, C., & Meier, A. (2016). Are we sustainable? Promoting a culture of sustainability in planned communities with a sustainability focus. In *QScience Proceedings, Qatar Green Building Conference 2016*. The Action, Nov 2016 (Vol. 15). https://doi.org/10.5339/qproc.2016.qgbc.15

Diaz, R., Selman, M., Chique, C. (2011). Global eutrophic and hypoxic coastal systems. World Resources Institute. Eutrophication and hypoxia: Nutrient pollution in coastal waters.

Diop, S., & Asongu, S. (2021). Research productivity: Trend and comparative analyses by regions and continents. *European Xtramile Centre of African Studies*. https://doi.org/10.2139/ssrn.3855361

Duhec, A. V., Jeanne, R. F., Maximenko, N., & Hafner, J. (2015). Composition and potential origin of marine debris stranded in the Western Indian Ocean on remote Alphonse Island, Seychelles. *Marine Pollution Bulletin, 96*, 76–86. https://doi.org/10.1016/j.marpolbul.2015.05.042

EPA (2020). *Best practices for solid waste management: A guide for decision-makers in developing countries* (EPA 530-R-20-002). https://www.epa.gov/sites/default/files/2020-10/documents/master_swmg_10-20-20_0.pdf

Filho, W. L., Voronova, V., Kloga, M., Paço, A., Minhas, A., Salvia, A. L., et al. (2021). COVID-19 and waste production in households: A trend analysis. *Science of the Total Environment, 777*, 145997. https://doi.org/10.1016/j.scitotenv.2021.145997

Gordon, S. (2021). Mass preferences for the free movement of people in Africa: A public opinion analysis of 36 countries. *International Migration Review, 56*, 1. https://doi.org/10.1177/019791 83211026243

Grant, R., & Oteng-Ababio, M. (2016). The global transformation of materials and the emergence of informal "Urban Mining" in Accra, Ghana. *Africa Today, 62*, 2–20. https://doi.org/10.2979/africatoday.62.4.01

GRID-Arendal. (2020). *Proceedings of the workshop on preventing and managing Marine Litter in West, Central and Southern Africa.*

Haider, N., Osman, A. Y., Gadzekpo, A., Akipede, G. O., Asogun, D., Ansumana, R., Lessells, R. J., Khan, P., Hamid, M. M. A., Yeboah-Manu, D., & Mboera, L. (2020). Lockdown measures in response to COVID-19 in nine sub-Saharan African countries. *BMJ Global Health, 5*. https://doi.org/10.1136/bmjgh-2020-003319

Inegbedion, H. (2021). Impact of COVID-19 on economic growth in Nigeria: Opinions and attitudes. *Heliyon 7*. https://doi.org/10.1016/j.heliyon.2021.e06943

International Monetary Fund. (2021). *IMF Datamapper: Real GDP growth.* https://www.imf.org/external/datamapper/NGDP_RPCH@WEO/OEMDC/ADVEC/WEOWORLD/AFQ. Accessed Febraury 25, 21.

IUCN, EA, QUANTIS. (2020a). *National guidance for plastic pollution hotspotting and shaping action, country report report Kenya.* United Nations Environment Programme. https://plasticho tspotting.lifecycleinitiative.org/wp-content/uploads/2020a/12/kenya_final_report_2020a.pdf

IUCN, EA, QUANTIS. (2020b). *National guidance for plastic pollution hotspotting and shaping action, country report Mozambique.* United Nations Environment Programme. https://plasti chotspotting.lifecycleinitiative.org/wp-content/uploads/2020b/12/mozambique_final_report_ 2020b.pdf

IUCN, EA, QUANTIS. (2021a). *National guidance for plastic pollution hotspotting and shaping action, country report Tanzania.* https://www.iucn.org/sites/dev/files/content/documents/tan zania_-_national_guidance_for_plastic_pollution_hotspotting_and_shaping_action_-_2021a. pdf

IUCN, EA, QUANTIS, (2021b). *National guidance for plastic pollution hotspotting and shaping action, country report report South Africa.* United Nations Environment Programme. https:// www.iucn.org/sites/dev/files/content/documents/south_africa_-_national_guidance_for_pla stic_pollution_hotspotting_and_shaping_action_-_2021b.pdf

Jambeck, J., Hardesty, B. D., Brooks, A.L., Friend, T., Teleki, K., Fabres, J., Beaudoin, Y., Bamba, A., Francis, J., Ribbink, A. J., & Baleta, T. (2018). Challenges and emerging solutions to the land-based plastic waste issue in Africa. *Marine Policy, 96*, 256–263. https://doi.org/10.1016/j. marpol.2017.10.041

Lyare, P. U., Ouki, S. K., Bond, T. (2020). Microplastics removal in wastewater treatmentplants: A critical review. *Environmental Science: Water Research & Technology, 6*, 2664. https://doi.org/ 10.1039/D0EW00397B

Maes, T., & Preston-Whyte, F. (2022). E-waste it wisely: Lessons from Africa. *SN Applied Sciences, 4*, 72. https://doi.org/10.1007/s42452-022-04962-9

Malone, T. C., & Newton, A. (2020). The globalization of cultural eutrophication in the coastal ocean: Causes and consequences. *Frontiers in Marine Science, 7*. https://doi.org/10.3389/fmars. 2020.00670

Odeyingbo, O., Nnorom, I. C., & Deubzer, O., (2017). *Person in the port project: Assessing import of used electrical and electronic equipment into Nigeria.* Bonn.

Okoffo, E. D., O'Brien, S., Ribeiro, F., Burrows, S. D., Toapanta, T., Rauert, C., O'Brien, J. W., Tscharke, B. J., Wang, X., & Thomas, K. V. (2021). Plastic particles in soil: State of the knowledge on sources, occurrence and distribution, analytical methods and ecological impacts.

Environmental Science-Processes & Impacts, 23, 240–274. https://doi.org/10.1039/D0EM00
312C

Okuku, E., Kiteresi, L., Owato, G., Otieno, K., Mwalugha, C., Mbuche, M., Gwada, B., Nelson,
A., Chepkemboi, P., Achieng, Q., & Wanjeri, V. (2021a). The impacts of COVID-19 pandemic
on marine litter pollution along the Kenyan Coast: A synthesis after 100 days following the first
reported case in Kenya. *Marine Pollution Bulletin, 162,* 111840. https://doi.org/10.1016%2Fj.
marpolbul.2020.111840

Prata, J. C., Silva, A. L. P., Walker, T. R., Duarte, A. C., & Rocha-santos, T. (2020). COVID-
19 pandemic repercussions on the use and management of plastics. *Environmental Science and
Technology, 54,* 7760–7765. https://doi.org/10.1021/acs.est.0c02178

Ryan, P. G. (2020a). Land or sea? What bottles tell us about the origins of beach litter in Kenya.
Waste Management, 116, 49–57. https://doi.org/10.1016/j.wasman.2020.07.044

Ryan, P. G., (2020b). The transport and fate of marine plastics in South Africa and adjacent oceans.
South African Journal of Science, 116, 5–6. https://doi.org/10.17159/sajs.2020b/7677

Ryan, P. G., Perold, V., Osborne, A., & Moloney, C. L. (2018). Consistent patterns of debris on
South African beaches indicate that industrial pellets and other mesoplastic items mostly derive
from local sources. *Environmental Pollution.* https://doi.org/10.1016/j.envpol.2018.02.017

Ryan, P. G., Maclean, K., & Weideman, E. A. (2020a). The impact of the COVID-19 lockdown
on urban street litter in South Africa. *Environmental Processes, 7,* 1302–1312. https://doi.org/0.
1007/s40710-020-00472-1

Ryan, P. G, Pichegru, L., Perold, V., & Moloney, C. L. (2020b). Monitoring marine plastics-will we
know if we are making a difference? *South African Journal of Science, 116,* 7678. https://doi.
org/10.17159/sajs.2020b/7678

Ryan, P. G., Weideman, E. A., Perold, V., Hofmeyr, G., & Connan, M. (2021). Message in a
bottle_Assessing the sources and origins of beach litter to tackle marine pollution. *Enviornmental
Justice, 288.* https://doi.org/10.1016/j.envpol.2021.117729

Scarlat, N., Motola, V., Dallemand, J. F., Monforti-Ferrario, F., & Mofor, L. (2015). Evaluation of
energy potential of municipal solid waste from African urban areas. *Renewable and Sustainable
Energy Reviews, 50,* 1269–1286. https://doi.org/10.1016/j.rser.2015.05.067

The Global Economy. (2021). *Research and development expenditure—Country rankings.* https://
www.theglobaleconomy.com/rankings/Research_and_development/. Accessed 17 August, 21.

UNCLOS. (1986). *United Nations convention on the law of the sea.* https://doi.org/10.18356/f80
44229-en

UNEP. (2018). *Africa waste management outlook.* Nairobi, Kenya. https://wedocs.unep.org/handle/
20.500.11822/25514

United Nations. (2020). *Economic report on Africa 2020: Innovative finance for private sector
development in Africa development in Africa.* Addis Ababa, Ethiopia. https://repository.uneca.
org/handle/10855/43834

United Nations. (2022). *World economic situation and prospects 2022.* New York. https://www.un.
org/development/desa/dpad/wp-content/uploads/sites/45/publication/WESP2022_web.pdf

Van Der Mheen, M., Van Sebille, E., & Pattiaratchi, C. (2020). Beaching patterns of plastic debris
along the Indian Ocean rim. *Ocean Science, 16,* 1317–1336. https://doi.org/10.5194/os-16-1317-
2020

Weideman, E. A., Perold, V., Omardien, A., Smyth, L. K., & Ryan, P. G. (2020). Quantifying
temporal trends in anthropogenic litter in a rocky intertidal habitat. *Marine Pollution Bulletin,
160,* 10. https://doi.org/10.1016/j.marpolbul.2020.111543

Wilson, D. C. (1996). Stick or carrot? The use of policy measures to move waste management up
the hierarchy. *Waste Management & Research, 14,* 4. https://doi.org/10.1006/wmre.1996.0039

Wilson, D. C., Rodic, L., Modak, P., Soos, R., Carpintero, A., Velis, K., Iyer, M., & Simonett, O. (2015). *Global waste management outlook*. International Solid Waste Association and United National Environment Programme. https://www.uncclearn.org/wp-content/uploads/library/une p23092015.pdf

Yasin, J. A., Kroeze, C., & Mayorga, E. (2010). Nutrients export by rivers to the coastal waters of Africa: Past and future trends. *Global Biogeochemical Cycles, 24*. https://doi.org/10.1029/200 9GB003568

Printed in the United States
by Baker & Taylor Publisher Services